Lecture Notes in Mathematics

C.I.M.E. Foundation Subseries

Volume 2283

More information about this subseries at http://www.springer.com/series/3114

Fondazione C.I.M.E., Firenze

C.I.M.E. stands for *Centro Internazionale Matematico Estivo*, that is, International Mathematical Summer Centre. Conceived in the early fifties, it was born in 1954 in Florence, Italy, and welcomed by the world mathematical community: it continues successfully, year for year, to this day.

Many mathematicians from all over the world have been involved in a way or another in C.I.M.E.'s activities over the years. The main purpose and mode of functioning of the Centre may be summarised as follows: every year, during the summer, sessions on different themes from pure and applied mathematics are offered by application to mathematicians from all countries. A Session is generally based on three or four main courses given by specialists of international renown, plus a certain number of seminars, and is held in an attractive rural location in Italy.

The aim of a C.I.M.E. session is to bring to the attention of younger researchers the origins, development, and perspectives of some very active branch of mathematical research. The topics of the courses are generally of international resonance. The full immersion atmosphere of the courses and the daily exchange among participants are thus an initiation to international collaboration in mathematical research.

C.I.M.E. Director (2002 – 2014)
Pietro Zecca
Dipartimento di Energetica "S. Stecco"
Università di Firenze
Via S. Marta, 3
50139 Florence
Italy
e-mail: zecca@unifi.it

C.I.M.E. Director (2015 –)
Elvira Mascolo
Dipartimento di Matematica "U. Dini"
Università di Firenze
viale G.B. Morgagni 67/A
50134 Florence
Italy
e-mail: mascolo@math.unifi.it

C.I.M.E. Secretary
Paolo Salani
Dipartimento di Matematica "U. Dini"
Università di Firenze
viale G.B. Morgagni 67/A
50134 Florence
Italy
e-mail: salani@math.unifi.it

CIME activity is carried out with the collaboration and financial support of INdAM (Istituto Nazionale di Alta Matematica)

For more information see CIME's homepage: **http://www.cime.unifi.it**

Claudia Polini • Claudiu Raicu • Matteo Varbaro •
Mark E. Walker

Recent Developments in Commutative Algebra

Levico Terme, Trento 2019

Aldo Conca • Srikanth B. Iyengar • Anurag K. Singh
Editors

In collaboration with

Authors
Claudia Polini
Department of Mathematics
University of Notre Dame
Notre Dame, IN, USA

Claudiu Raicu
Department of Mathematics
University of Notre Dame
Notre Dame, IN, USA

Matteo Varbaro
Dipartimento di Matematica
Università di Genova
Genova, Italy

Mark E. Walker
Department of Mathematics
University of Nebraska–Lincoln
Lincoln, NE, USA

Editors
Aldo Conca
Dipartimento di Matematica
Università di Genova
Genova, Italy

Srikanth B. Iyengar
Department of Mathematics
University of Utah
Salt Lake City, UT, USA

Anurag K. Singh
Department of Mathematics
University of Utah
Salt Lake City, UT, USA

ISSN 0075-8434   ISSN 1617-9692 (electronic)
Lecture Notes in Mathematics
C.I.M.E. Foundation Subseries
ISBN 978-3-030-65063-6   ISBN 978-3-030-65064-3   (eBook)
https://doi.org/10.1007/978-3-030-65064-3

Mathematics Subject Classification: Primary: 13-02; Secondary: 13A30, 13D02; 14H51, 13P10, 13D15, 19L20

This Springer imprint is published by the registered company Springer Nature Switzerland AG.
The registered company address is: Gewerbestrasse 11, 6330 Cham, Switzerland

Preface

The CIME-CIRM course *Recent Developments in Commutative Algebra* took place in Levico Terme, Trento, July 1–5, 2019. Commutative Algebra has witnessed a number of spectacular developments over the last few years, including the resolution of long-standing problems; the new techniques and perspectives are leading to an extraordinary transformation in the field. The workshop presented some of these developments, with the aim of spurring further advances. The four chapters that follow are based on the lectures delivered by four speakers.

The Rees ring and the special fiber ring of an ideal arise in the process of blowing up a variety along a subvariety. Rees rings and special fiber rings also describe, respectively, the graphs and the images of rational maps between projective spaces. A difficult open problem in commutative algebra, algebraic geometry, elimination theory, and geometric modeling is to determine explicitly the equations defining graphs and images of rational maps. Claudia Polini's lectures discuss various aspects of these questions, including their solution for ideals of low codimensions.

Claudiu Raicu introduces the theory of Koszul modules and surveys some applications to syzygies and to Green's conjecture, following the recent work of Aprodu, Farkas, Papadima, Raicu, and Weyman. He presents a relationship between Koszul modules and the syzygies of the tangent developable surface to a rational normal curve, which arises from a version of Hermite reciprocity for SL_2-representations. The lectures go on to describe, in terms of the characteristic of the underlying field, which of the Betti numbers of this surface are zero and which ones are not. By passing to a hyperplane section and using degenerations, the generic version of Green's conjecture can be deduced in almost all characteristics.

Matteo Varbaro begins with an introduction to the theory of Gröbner basis and Buchberger's algorithm, with a special focus on deformation aspects and Gröbner degenerations. The goal of his lectures is to present a self-contained proof of a recent result, obtained in a joint work with Conca: if a homogenous ideal in a polynomial ring admits a monomial order such that the initial ideal is generated by square-free monomials, then the local cohomology modules of the original ideal and the initial ideal have the same Hilbert function; this implies, in particular, that the ideals have the same depth and regularity and settles a conjecture of Herzog.

In their original form, Adams operations refer to natural operators that are defined on the Grothendieck group of algebraic vector bundles on an algebraic variety, or on the Grothendieck group of topological vector bundles on a topological space. These operations determine a decomposition of the Grothendieck groups (modulo torsion) into the Chow groups on the algebraic side, and the singular cohomology groups on the topological side. Variants of these classical operators have been defined in a local algebraic setting as well. For example, in the 1980s, Gillet and Soule used a notion of Adams operations defined on certain Grothendieck groups of bounded complexes of free modules over a regular local ring in their proof of Serre's Vanishing Conjecture. In his lectures, Mark Walker explains these applications, as well as his own work on the total rank conjecture, a variation of the Buchsbaum-Eisenbud-Horrocks conjecture.

The workshop was attended by 51 participants, including 43 Ph.D. students or early career mathematicians; we are very grateful to them for the enthusiasm and the energy that went a long way toward making this a successful event. A special note of thanks to Alessio D'Alì, Janina Carmen Letz, Devlin Mallory, and Alessio Sammartano, for their assistance with proofreading the lecture notes.

Lastly, we thank the CIME and the CIRM center for supporting the school; in particular, we are very grateful to Elvira Mascolo and Paolo Salani of CIME, and to Marco Andreatta and Augusto Micheletti of CIRM.

Genova, Italy
Salt Lake City, UT, USA
Salt Lake City, UT, USA
September 2020

Aldo Conca
Srikanth B. Iyengar
Anurag K. Singh

Participants

- Ayah Almousa, Cornell University
- Nasrin Altafi, KTH Stockholm
- Taylor Ball, University of Notre Dame
- Rémi Bignalet-Cazalet, Universitá di Genova
- Alessio Caminata, Université de Neuchâtel
- Aldo Conca, Universitá di Genova
- Laura Cossu, Universitá di Padova
- Alessio D'Alì, University of Warwick
- Alessandro De Stefani, Universitá di Genova
- Peng Du, Université Grenoble Alpes
- Zachary Flores, Colorado State University
- Zachary Greif, Iowa State University
- Lorenzo Guerrieri, Universitá di Catania
- Roser Homs Pons, Universitat de Barcelona
- Srikanth Iyengar, University of Utah
- Yeongrak Kim, Universität des Saarlandes
- Mitra Koley, Chennai Mathematical Institute
- Janina Carmen Letz, University of Utah
- Alessandra Maria Ausilia Licata, Universitá di Catania
- Devlin Mallory, University of Michigan
- Amadeus Martin, University of Nebraska
- Carla Mascia, Universitá di Trento
- Matthew Mastroeni, Oklahoma State University
- Nicola Maugeri, Universitá di Catania
- Luca Mesiti, Universitá di Genova
- Tsutomu Nakamura, Universitá di Verona
- Navid Nemati, Sorbonne Université
- Lisa Nicklasson, Stockholms Universitet
- Milo Orlich, Aalto University
- Janet Page, University of Bristol
- Claudia Polini, University of Notre Dame
- Josh Pollitz, University of Nebraska
- Claudiu Raicu, University of Notre Dame
- Giancarlo Rinaldo, Universitá di Trento
- Francesco Romeo, Universitá di Trento
- Kumari Saloni, Chennai Mathematical Institute
- Alessio Sammartano, University of Notre Dame
- Lisa Seccia, Universitá di Genova
- Anurag Singh, University of Utah
- Dumitru Stamate, Universitatea din Bucureşti
- Jonathan Steinbuch, Universität Osnabrück
- Francesco Strazzanti, Universitá di Catania

- Janet Striuli, Fairfield University
- Danny Troia, Universitá di Catania
- Van Duc Trung, Universitá di Genova
- Manolis Tsakiris, ShanghaiTech University
- Matteo Varbaro, Universitá di Genova
- Lorenzo Venturello, Universität Osnabrück
- Ulrich von der Ohe, Universitá di Genova
- Mark Walker, University of Nebraska
- Hongmiao Yu, Universitá di Genova

Contents

List of Contributors

Claudia Polini Department of Mathematics, University of Notre Dame, Notre Dame, IN, USA

Claudiu Raicu Department of Mathematics, University of Notre Dame, Notre Dame, IN, USA
Institute of Mathematics "Simion Stoilow" of the Romanian Academy, Bucharest, Romania

Matteo Varbaro Dipartimento di Matematica, Università di Genova, Genova, Italy

Mark E. Walker Department of Mathematics, University of Nebraska-Lincoln, Lincoln, NE, USA

Chapter 1
Defining Equations of Blowup Algebras

Claudia Polini

Abstract This is the content of four lectures on Rees rings and their defining equations delivered at the CIME workshop *Recent developments in Commutative Algebra* held at Levico Terme (Trento), Italy on July 1–5, 2019. The Rees ring and the special fiber ring of an ideal arise in the process of blowing up a variety along a subvariety. Rees rings and special fiber rings also describe, respectively, the graphs and the images of rational maps between projective spaces. A difficult open problem in commutative algebra, algebraic geometry, elimination theory, and geometric modeling is to determine explicitly the equations defining graphs and images of rational maps. In these lectures we will discuss this topic in several situations.

1.1 Introduction

Blowup algebras are ubiquitous in Commutative Algebra and Algebraic Geometry. Below we give an incomplete list of research areas where they play an important role. Hopefully this list will motivate the reader to study blowup algebras and learn to love them.

1. *Resolution of singularities.* Blowup algebras are the algebras associated to the blow-up of a variety along a subvariety.
2. *Macaulayfication.* Macaulayfication is a mild version of resolution of singularities: the new scheme is only required to be Cohen-Macaulay rather than smooth.
3. *Multiplicity theory.* The notions of Hilbert function, Hilbert polynomial, and Hilbert series are generalized to local rings using blowup algebras.
4. *Generalized multiplicities.* Using blowup algebras, general notions of multiplicity are defined for arbitrary submodules of a free module.

C. Polini (✉)
Department of Mathematics, University of Notre Dame, Notre Dame, IN, USA
e-mail: cpolini@nd.edu

A. Conca et al. (eds.), *Recent Developments in Commutative Algebra*, Lecture
Notes in Mathematics 2283, https://doi.org/10.1007/978-3-030-65064-3_1

5. *Asymptotic properties of ideals.* Blowup algebras encode all powers of an ideal (in general all members of a filtration). For instance, the Hilbert function of the special fiber ring encodes the minimal number of generators of all the powers of an ideal in a local ring.

6. *Integral dependence.* Blowup algebras are the environment where integral dependence of ideals and modules takes place. Questions like *how many powers of an ideal need to be tested in order to guarantee that all powers are integrally closed* can be formulated in terms of Rees algebras.

7. *Reductions, Briançon-Skoda theorem, cores, and multiplier ideals.* The notion of reduction is best defined using Rees algebras. Reductions are simplifications of a given ideal. Since they are highly non-unique, one considers their intersection, called the core of an ideal, which encodes information about all possible reductions. The Briançon-Skoda theorem says that if I is an ideal of a regular local ring of dimension d then the integral closure of I^d is contained in the core of I.

8. *Equisingularity theory.* The goal in equisingularity theory is to devise criteria for analytic sets that occur in a flat family to be alike. Many of these criteria are based on multiplicities and are proved via Rees algebras of ideals and modules.

9. *Symbolic powers and set-theoretic complete intersections.* The Noetherianness of the symbolic Rees algebra of an ideal I is related to the number of equations that are needed to define the subvariety $V(I)$ set-theoretically.

10. *Images and graphs of rational maps between projective spaces.* Images and graphs of rational maps between projective spaces are described algebraically by special fiber rings and Rees rings.

11. *Defining equations of Segre products, Veronese embeddings, Gauss images, tangential and secant varieties.* All these varieties are given by Rees algebras and special fiber rings.

12. *Geometric modeling.* The problem of describing the implicit equations of curves and surfaces given parametrically has important applications in geometric modeling and in computer-aided-design, where it is known as the implicitization problem.

13. *Singularities of rational plane curves.* A rational curve is the image of a morphism. There is a correspondence between the constellations of singularities on or infinitely near the curve and numerical features of the Rees algebra of the ideal generated by the forms defining the morphism.

14. *Eisenbud-Goto conjecture.* The counterexample to the Eisenbud-Goto conjecture was constructed using Rees like algebras.

Now we turn to the definition of blowup algebras.

Definition 1.1 Let R be a Noetherian ring (often we will assume that $(R, \mathfrak{m})$ is local or positively graded over a local ring with $\mathfrak{m}$ its homogenous maximal ideal). Let I be an R-ideal (if R is graded we will always assume that I is homogeneous). The blowup algebras of I are the *Rees algebra* $\mathcal{R}(I) = R[It]$, the *associated graded ring* $\mathcal{G}(I) = R[It] \otimes_R R/I$, and the *special fiber ring* $\mathcal{F}(I) = R[It] \otimes_R R/\mathfrak{m}$.

These algebras are best studied via the *symmetric algebra* $S(I)$:

$$S(I) = \bigoplus_{j=0}^{\infty} \operatorname{Sym}_j(I)$$

$$\downarrow$$

$$\mathcal{R}(I) = R[It] = \bigoplus_{j=0}^{\infty} I^j t^j \subset R[t]$$

$$\downarrow$$

$$\mathcal{G}(I) = R[It] \otimes_R R/I = \bigoplus_{j=0}^{\infty} I^j / I^{j+1}$$

$$\downarrow$$

$$\mathcal{F}(I) = R[It] \otimes_R R/\mathfrak{m} = \bigoplus_{j=0}^{\infty} I^j / \mathfrak{m} I^j$$

where, for the last map, I is assumed to be a proper ideal. The projective spectrum of the Rees ring is the blow-up of $\operatorname{Spec}(R)$ along $V(I)$, the projective spectrum of the associated graded ring is the exceptional fiber of the blow-up, and the projective spectrum of the special fiber ring is the special fiber of the blow-up.

The study of blowup rings has been a central problem in commutative algebra since the seventies. Among others, the questions that have been investigated are

1. When are blowup algebras Cohen-Macaulay?
2. How are properties of these algebras, such as their Cohen-Macaulayness, related?
3. What is the canonical module of blowup algebras, and when are these algebras Gorenstein?
4. When are blowup algebras normal?
5. What are their defining equations?
6. Often it is difficult to compute explicitly the defining equations. An important step is to bound their degrees.
7. A rational curve is the image of a morphism. To understand the image it is often better to study the graph of the morphism since it encodes more information than the image itself. The special fiber ring is the homogenous coordinate ring of the image and the Rees ring is the bihomogenous coordinate ring of the graph. An important problem is to relate numerical invariants of the Rees ring, such as the bidegrees of its defining equations, to local properties of the rational curve, such as the constellation of its singularities.

In these notes we will mainly focus on problems (5) and (6). We begin by recording the Krull dimension of blowup algebras.

Theorem 1.1 *Assume that I is a proper ideal of a local ring $(R, \mathfrak{m})$ of dimension d. Then the dimension of the symmetric algebra is the Foster number,*

$$\dim S(I) = \sup\{\mu(I_{\mathfrak{p}}) + \dim(R/\mathfrak{p}) \mid \mathfrak{p} \in \mathrm{Spec}(R)\},$$

where μ denotes the minimal number of generators.

If I has positive height, then the dimension of $\mathcal{R}(I)$ is $d+1$, the dimension of $\mathcal{G}(I)$ is d, and the dimension of $\mathcal{F}(I)$ is the analytic spread, $\ell(I)$, an integer satisfying

$$\ell(I) \leq d.$$

Proof The statement about the symmetric algebra is due to Huneke and Rossi [23]. The proofs about the dimensions of $\mathcal{R}(I)$ and $\mathcal{G}(I)$ can be found in [42]. The upper bound on the analytic spread is clear since $\mathcal{F}(I)$ is an epimorphic image of $\mathcal{G}(I)$. □

1.2 Linear Type

Let I be an ideal in a Noetherian ring R generated by $f_1, \ldots, f_n$. We map the polynomial ring $S = R[y_1, \ldots, y_n]$ onto $\mathcal{R}(I)$ by sending y_j to $f_j t$, and we denote by $\mathcal{J}$ the kernel of this map, which is the defining ideal of $\mathcal{R}(I)$. The map factors through the symmetric algebra $S(I)$. The defining ideal $\mathcal{L}$ of the symmetric algebra is readily available from a presentation matrix φ of $f_1, \ldots, f_n$ as we will explain next. Hence, to understand $\mathcal{J}$, it suffices to determine the kernel of the map $\alpha \colon S(I) \twoheadrightarrow \mathcal{R}(I)$, which is the R-torsion of the symmetric algebra when grade $I > 0$.

To describe the defining ideal of the symmetric algebra, we recall that the symmetric algebra $S(E)$ of any module E can be defined via the universal property: *for any R-linear map ψ from E to a commutative R-algebra A, there exists a unique R-algebra homomorphism $\overline{\psi}$ from $S(E)$ to A such that $\overline{\psi}|_E = \psi$.*

If E is a finite R-module with presentation

$$0 \longrightarrow Z \longrightarrow F = Ry_1 \oplus \cdots \oplus Ry_n \longrightarrow E \longrightarrow 0,$$

then, using the universal property, one sees that $S(F) = R[y_1, \ldots, y_n]$ and that there is an exact sequence

$$0 \longrightarrow Z \cdot S(F) \longrightarrow S(F) = R[y_1, \ldots, y_n] \longrightarrow S(E) \longrightarrow 0.$$

In particular, if Z is the column space of a matrix φ, the ideal $Z \cdot S(F)$ is generated by the entries of the row vector $[y_1, \ldots, y_n] \cdot \varphi$, that is

$$S(E) \cong R[y_1, \ldots, y_n]/([y_1, \ldots, y_n] \cdot \varphi).$$

The converse is also true: any quotient of a polynomial ring in the variables $y_1, \ldots, y_n$ by an ideal generated by linear forms is the symmetric algebra of an R-module.

As the equations of the symmetric algebra are well understood, the best-case scenario is when the Rees algebra and the symmetric algebra are isomorphic. In this case we say that the ideal I is of linear type:

Definition 1.2 An ideal I is *of linear type* if $S(I)$ and $\mathcal{R}(I)$ are isomorphic via the natural projection α, that is if $\mathcal{J} = \mathcal{L}$.

Next we give some necessary and sufficient conditions for an ideal I to be of linear type. We first show that the property of being of linear type is quite restrictive. Indeed an ideal of linear type is G_∞. In particular any ideal of linear type has at most d generators, where d is the dimension of the ring R.

Definition 1.3 An ideal I is G_∞ if $\mu(I_{\mathfrak{p}}) \leq \dim R_{\mathfrak{p}}$ for every prime ideal $\mathfrak{p} \in V(I)$.

Now assume that I is an ideal of linear type. If $(R, \mathfrak{m}, k)$ is a local ring, then the special fiber ring of I is isomorphic to the symmetric algebra of $I \otimes_R k \cong k^{\mu(I)}$, which is a polynomial ring over k in $\mu(I)$ variables. Indeed,

$$\mathcal{F}(I) = \mathcal{R}(I) \otimes_R k \cong S(I) \otimes_R k ,$$

where the isomorphism holds since I is of linear type. Now

$$S(I) \otimes_R k = S(I \otimes_R k) = \mathrm{Sym}_k(k^{\mu(I)}) .$$

Computing dimensions we obtain

$$d \geq \ell(I) = \mu(I) .$$

Hence in any Noetherian ring R we have

$$\mu(I_{\mathfrak{p}}) \leq \dim R_{\mathfrak{p}}$$

for every prime ideal $\mathfrak{p} \in V(I)$, as can be easily seen by applying the above fact to the local ring $R_{\mathfrak{p}}$. Thus we have proven:

Theorem 1.2 *An ideal of linear type is G_∞.*

The property of being G_∞ can be expressed in terms of the heights of Fitting ideals of I.

Exercise 1.1 Assume that I has positive height. The ideal I is G_∞ if and only if $\mathrm{ht}\, \mathrm{Fitt}_j(I) \geq j + 1$ for all $j \geq 1$.

Unfortunately the condition G_∞ is only necessary but not sufficient for an ideal to be of linear type.

Example 1.1 Let $R = k[x, y, z, w]$ be a polynomial ring over a field k. The ideal $I = (x, y) \cap (z, w)$ is G_∞ but is not of linear type.

What about sufficient conditions? At least if the grade of I is positive, the ideal being of linear type is equivalent to the symmetric algebra being torsion-free over R. To be able to establish such a property it would be useful to have a free resolution of the symmetric algebra. In general such resolutions are hard to find, however there are complexes called *approximation complexes* that are candidates for "resolutions" of the blowup rings, not by free modules, but by sufficiently 'nice' modules. The approximations complexes were first introduced by Simis and Vasconcelos in [47, 48] to study symmetric algebras, and they were further developed by Herzog-Simis-Vasconcelos [15–17].

Let I be an ideal generated by $f_1, \ldots, f_n$ in a Noetherian ring R. Let φ be a presentation matrix of $f_1, \ldots, f_n$, and let Z be the module of syzygies of $f_1, \ldots, f_n$,

$$0 \longrightarrow Z \longrightarrow F = Ry_1 \oplus \cdots \oplus Ry_n \longrightarrow I \longrightarrow 0,$$

where the last map sends each basis element y_j of F to the generator f_j of I.

Write $S = S(F) = R[y_1, \ldots, y_n]$. As seen earlier, we have the exact sequence

$$0 \longrightarrow Z \cdot S \longrightarrow S \longrightarrow S(I) \longrightarrow 0$$

which gives a homogeneous S-linear map $\widetilde{\varphi}$

$$S \otimes_R Z(-1) \overset{\widetilde{\varphi}}{\longrightarrow} S \longrightarrow S(I) \longrightarrow 0.$$

We consider the Koszul complex $K_\bullet(\widetilde{\varphi})$ as below:

$$\longrightarrow S \otimes_R \wedge^j Z(-j) \longrightarrow S \otimes_R \wedge^{j-1} Z(-j+1) \longrightarrow \cdots \longrightarrow S \otimes_R Z(-1) \overset{\widetilde{\varphi}}{\longrightarrow} S$$

with the usual differentials induced by $\widetilde{\varphi}$. This is a homogeneous complex of S-modules whose zeroth homology is the symmetric algebra of I:

$$H_0(K_\bullet(\widetilde{\varphi})) = S(I)$$

This complex is too long to be exact and to give good depth estimates on the symmetric algebra of I. However, $\mathrm{rank}_R Z = n - 1$, hence $\wedge^j Z$ is torsion for all $j > n - 1$. Thus, if we factor out the torsion, or even pass to the double dual, we obtain a complex of length $n-1$ that has a chance to be exact. The resulting complex looks like

$$0 \longrightarrow S \otimes_R (\wedge^{n-1} Z)^{**}(-n+1) \longrightarrow \ldots \longrightarrow S \otimes_R (\wedge^j Z)^{**}(-j) \longrightarrow \cdots$$
$$\ldots \longrightarrow S \otimes_R Z^{**}(-1) \longrightarrow S$$

where $*$ denotes the dual into R, that is $\mathrm{Hom}_R(-, R)$.

From the last complex we build yet another one. Notice that if the grade of I is at least two, then the modules $(\wedge^j Z)^{**}$ are naturally isomorphic to the Koszul cycles Z_j. Indeed, since the Koszul complex $K_\bullet(f_1, \ldots, f_n; R)$ on $f_1, \ldots, f_n$ has an algebra structure, there is a natural map from the module $\wedge^j Z$ to the Z_j. This natural map becomes an isomorphism after taking double duals provided the grade of I is at least two.

Lemma 1.1 *Let Z_i be the Koszul cycles of $f_1, \ldots, f_n$. If the ideal generated by $f_1, \ldots, f_n$ has grade at least two, then the algebra structure of $K_\bullet(f_1, \ldots, f_n; R)$ induces isomorphisms*

$$(\wedge^j Z)^{**} \xrightarrow{\sim} Z_j$$

for every j.

Proof As $Z_\bullet$ is a graded commutative algebra, there exist R-linear maps

$$\psi_j \colon \wedge^j Z \longrightarrow Z_j.$$

Dualizing twice we obtain the R-linear maps $(\psi_j)^{**} \colon (\wedge^j Z)^{**} \longrightarrow (Z_j)^{**}$. We consider also the natural R-linear maps $\phi_j \colon Z_j \longrightarrow (Z_j)^{**}$. We claim that the maps $(\psi_j)^{**}$ and ϕ_j are both isomorphisms.

Let $\mathfrak{p} \in \mathrm{Spec}(R)$ with depth $R_\mathfrak{p} \geq 2$. The right exact sequence

$$0 \longrightarrow (Z_j)_\mathfrak{p} \longrightarrow (K_j)_\mathfrak{p} \longrightarrow (K_{j-1})_\mathfrak{p},$$

where K_j are the modules in the Koszul complex $K_\bullet(f_1, \ldots, f_n; R)$, shows that the depth of $(Z_j)_\mathfrak{p}$ is at least two. Hence all the modules Z_j, $(Z_j)^{**}$, $(\wedge^j Z)^{**}$ have the property that their depth locally at $\mathfrak{p}$ is at least two whenever depth $R_\mathfrak{p} \geq 2$. Thus, to show that $(\psi_j)^{**}$ and ϕ_j are isomorphisms, it suffices to show that they are isomorphisms locally at $\mathfrak{p} \in \mathrm{Spec}(R)$ with depth $R_\mathfrak{p} \leq 1$.

However, since the grade of $I = (f_1, \ldots, f_n)$ is at least two, if depth $R_\mathfrak{p} \leq 1$, then $I_\mathfrak{p} = R_\mathfrak{p}$. Hence the Koszul complex locally at $\mathfrak{p}$ is split exact. It follows that the maps $(\psi_j)_\mathfrak{p}$ are isomorphisms. Moreover, the modules $(Z_j)_\mathfrak{p}$ are free and therefore the maps $(\phi_j)_\mathfrak{p}$ are isomorphisms. $\square$

Thus, when the ideal I has grade at least two, the earlier complex has the form

$$0 \longrightarrow S \otimes_R Z_n(-n) \longrightarrow \ldots \longrightarrow S \otimes_R Z_j(-j) \longrightarrow \ldots \longrightarrow S \otimes_R Z(-1) \longrightarrow S.$$

This complex is called the $\mathcal{Z}$-*complex*. We consider the $\mathcal{Z}$-complex even when the grade of I is not at least two. In the general case though, the construction of the complex is different and we will explain it below. Notice that the zeroth homology of the $\mathcal{Z}$-complex is still the symmetric algebra of I, as $\mathcal{Z}_\bullet$ and $K_\bullet(\widetilde{\varphi})$ coincide in degrees one and zero:

$$H_0(\mathcal{Z}_\bullet) = S(I)$$

The above discussion motivates the general definition of the approximation complexes. We consider a double complex of graded S-modules. Its rows are the Koszul complex $K_\bullet(\underline{y}; S)$ on the sequence $\underline{y} = y_1, \ldots, y_n$ with coefficients in S and its columns are the Koszul complex $\overline{K}_\bullet(\underline{f}; R) \otimes_R S$ on the sequence $\underline{f} = f_1, \ldots, f_n$. The differentials anticommute, so this is a double complex.

If we take cycles, homology, or boundaries of the vertical differentials, we obtain three complexes, dubbed *approximation complexes*, the $\mathcal{Z}$-complex (Z for cycles), the $\mathcal{M}$-complex, and the $\mathcal{B}$-complex (B for boundaries)

$$\mathcal{Z}_\bullet(\underline{f}) : 0 \longrightarrow S \otimes_R Z_n(\underline{f})(-n) \longrightarrow \cdots \longrightarrow S \otimes_R Z_1(\underline{f})(-1) \longrightarrow S ,$$

$$\mathcal{M}_\bullet(\underline{f}) : 0 \longrightarrow S \otimes_R H_n(\underline{f})(-n) \longrightarrow \cdots \longrightarrow S \otimes_R H_1(\underline{f})(-1) \longrightarrow S \otimes R/I ,$$

$$\mathcal{B}_\bullet(\underline{f}) : 0 \longrightarrow S \otimes_R B_n(\underline{f})(-n) \longrightarrow \cdots \longrightarrow S \otimes_R B_1(\underline{f})(-1) \longrightarrow S \otimes I .$$

The differentials are induced by the differential on $K_\bullet(\underline{y}; S)$. Notice that the zeroth homology of $\mathcal{M}_\bullet(\underline{f})$ is the symmetric algebra of the conormal module I/I^2. The graded strands of the complexes are complexes of R-modules denoted by $\mathcal{Z}_{i\bullet}(\underline{f})$, $\mathcal{M}_{i\bullet}(\underline{f})$, $\mathcal{B}_{i\bullet}(\underline{f})$; for instance

$$\mathcal{Z}_{i\bullet}(\underline{f}) : 0 \longrightarrow \mathrm{Sym}_0(F) \otimes_R Z_i \longrightarrow \cdots \longrightarrow \mathrm{Sym}_{i-j}(I) \otimes_R Z_j \longrightarrow \cdots \longrightarrow \mathrm{Sym}_i(F)$$

with $H_0(\mathcal{Z}_{i\bullet}(\underline{f})) = \mathrm{Sym}_i(I) = S_i(I)$.

The approximation complexes depend on the choice of the generators of I, however their homology does not [48]. The properties of these complexes, in particular their acyclicity, are treated in detail in [15–17]. We want to describe when $\mathcal{Z}_\bullet(\underline{f})$ and $\mathcal{M}_\bullet(\underline{f})$ are acyclic, because in this case these complexes approximate, in a way, a free resolution of their zeroth homology $S(I)$ and $\mathrm{Sym}_{R/I}(I/I^2)$. Since the homology of $\mathcal{Z}_\bullet(\underline{f})$ and $\mathcal{M}_\bullet(\underline{f})$ does not depend on the $\underline{f}$ but only on I, we write $H_\bullet(\mathcal{Z}_\bullet(I))$ and $H_\bullet(\mathcal{M}_\bullet(I))$.

Proposition 1.1 *For any integer $i \neq -1$ there is a long exact sequence of homology*

$$\longrightarrow H_j(\mathcal{Z}_{i+1\bullet}(I)) \longrightarrow H_j(\mathcal{Z}_{i\bullet}(I)) \longrightarrow H_j(\mathcal{M}_{i\bullet}(I)) \longrightarrow H_{j-1}(\mathcal{Z}_{i+1\bullet}(I)) \longrightarrow$$

Proof We have exact sequences of complexes

$$0 \longrightarrow \mathcal{B}_\bullet(\underline{f}) \longrightarrow \mathcal{Z}_\bullet(\underline{f}) \longrightarrow \mathcal{M}_\bullet(\underline{f}) \longrightarrow 0 \tag{1.1}$$

$$0 \longrightarrow \mathcal{Z}_\bullet(\underline{f}) \longrightarrow K_\bullet(\underline{y}; S) \longrightarrow \mathcal{B}_\bullet(\underline{f})[-1](-1) \longrightarrow 0. \tag{1.2}$$

The long exact sequence of homology for (1.1) yields

$$\cdots \longrightarrow H_j(\mathcal{B}_\bullet(I)) \longrightarrow H_j(\mathcal{Z}_\bullet(I)) \longrightarrow H_j(\mathcal{M}_\bullet(I)) \longrightarrow H_{j-1}(\mathcal{B}_\bullet(I)) \longrightarrow \cdots$$

By (1.2), since the Koszul complex on $\underline{y}$ is acyclic, we obtain the isomorphism $H_j(\mathcal{B}_\bullet(I)) \cong H_j(\mathcal{Z}_\bullet(I))(1)$ for all $j \neq 0$ and the exact sequence

$$0 \longrightarrow H_0(\mathcal{B}_\bullet(I))(-1) \longrightarrow H_0(\mathcal{Z}_\bullet(I)) \longrightarrow R \longrightarrow 0$$

for $j = 0$. Hence $H_0(\mathcal{B}_{i\bullet}(I)) \cong H_0(\mathcal{Z}_{i+1\bullet}(I))$ as long as $i \neq -1$. $\qquad\square$

Notice that by the above proposition for every $i \geq 0$ there is an exact sequence

$$H_0(\mathcal{Z}_{i+1\bullet}(I)) = S_{i+1}(I) \xrightarrow{\lambda_i} H_0(\mathcal{Z}_{i\bullet}(I)) = S_i(I) \to H_0(\mathcal{M}_{i\bullet}(I)) = S_i(I/I^2) \to 0$$

and $\lambda_i(a_1 \cdot \ldots \cdot a_{i+1}) = a_1(a_2 \cdot \ldots \cdot a_{i+1})$, where $\cdot$ denotes the product in the symmetric algebra and $a_\nu \in I$. The homomorphism $\lambda_\bullet : [S(I)(1)]_{\geq 0} \longrightarrow S(I)$ is called the *downgrading homomorphism*. We obtain the following corollary:

Corollary 1.1 *The complex $\mathcal{M}_\bullet(\underline{f})$ is acyclic if and only if $\mathcal{Z}_\bullet(\underline{f})$ is acyclic and $\lambda_\bullet$ is injective. In addition, $\lambda_\bullet$ is injective if and only if I is of linear type.*

Thus the acyclicity of $\mathcal{M}_\bullet(\underline{f})$ will gives the isomorphism $S(I) \cong \mathcal{R}(I)$ and also a "resolution" of $\mathcal{R}(I)$ that can shed light on its Cohen-Macaulayness. We are left to investigate the question:

When is $\mathcal{M}_\bullet(\underline{f})$ acyclic?

Since the property G_∞ is necessary for I to be of linear type, we assume that I is G_∞. In this case the approximation complexes are short enough to apply the acyclicity lemma. We recall the acyclicity lemma, due to Peskine and Szpiro [45]:

Lemma 1.2 *Let*

$$C_\bullet : 0 \longrightarrow C_n \longrightarrow \cdots \longrightarrow C_0$$

be a complex of finite modules over a Noetherian local ring. Assume that for all $i > 0$

1. $H_j(C_\bullet) = 0$ or depth $H_j(C_\bullet) = 0$ and
2. depth $C_j \geq j$.

Then $C_\bullet$ is acyclic.

Thus to apply the acyclicity lemma to the complex $M_\bullet$ we need some conditions on the depth of the Koszul homology modules $H_j(\underline{f})$. We now assume that the ring R is local and Cohen-Macaulay of dimension d.

Definition 1.4 The ideal $I = (f_1, \ldots, f_n)$ has *sliding depth* if the homology modules of the Koszul complex $K_\bullet(\underline{f})$ satisfy

$$\text{depth } H_j(\underline{f}) \geq d - n + j.$$

The above property is independent of the choice of the generators of I and is preserved under localization at $\mathfrak{p} \in V(I)$ [15]. By Huneke [21], the sliding depth property is an invariant of even linkage. Therefore every licci ideal (meaning every ideal in the linkage class of a complete intersection) has sliding depth. In particular, perfect ideals of grade two and perfect Gorenstein ideals of grade three have sliding depth [1, 13, 53].

Now we are ready to state the main theorem of this section [15].

Theorem 1.3 (Herzog-Simis-Vasconcelos) *Let $I = (f_1, \ldots, f_n)$ be an ideal of positive height in a Cohen-Macaulay local ring. If I is G_∞ and has sliding depth, then $M_\bullet(\underline{f})$ is acyclic. In particular, I is of linear type and the blowup algebras are Cohen-Macaulay.*

We prove that each graded strand of $M_\bullet$ is acyclic. We localize at a minimal prime of $\text{Supp}_R\left(\bigoplus_{j>0} H_j(M_{t\bullet}(I))\right)$ to assume that $\text{Supp}_R\left(\bigoplus_{j>0} H_j(M_{t\bullet}(I))\right) \subset \{\mathfrak{m}\}$, so that the homology of this complex is either zero or has depth zero. Without changing the homology of the complex we may assume that $M_{t\bullet}(I) = M_{t\bullet}(\underline{f})$, where $\underline{f} = f_1, \ldots, f_n$ is a minimal generating set of I. Since sliding depth and G_∞ localize, the assumptions on the ideal are preserved. According to the acyclicity lemma, the strand $M_{t\bullet}(I)$ will be acyclic if the j-th module in the strand has depth at least j. The latter condition is verified using sliding depth and the G_∞ condition. Indeed, by G_∞, one has $d - n \geq 0$, hence the sliding depth condition implies that depth $H_j(\underline{f}) \geq j$.

The acyclicity of $M_\bullet$ implies that I is linear type and $Z_\bullet$ is acyclic. The Cohen-Macaulayness of $\mathcal{R}(I)$ is proved chasing depth along the $Z_\bullet$-complex. Indeed, the sliding depth condition gives depth $Z_j(\underline{f}) \geq d-n+j+1$. The Cohen-Macaulayness of $\mathcal{G}(I)$ follows from the Cohen-Macaulayness of $\mathcal{R}(I)$ according to [20].

We end this chapter with an important notion introduced by Huneke [19] to study powers of ideals.

Definition 1.5 A sequence $f_1, \ldots, f_n$ of elements of R is called a *d-sequence* if $(f_1, \ldots, f_i) : f_{i+1} f_k = (f_1, \ldots, f_i) : f_k$ for $0 \le i \le n - 1$ and $k \ge i + 1$.

The $\mathcal{M}$-complex is related to d-sequences very much like the Koszul complex is related to regular sequences. More specifically, when the residue field is infinite, the $\mathcal{M}$-complex is acyclic if and only if I can be generated by a d-sequence, and, every ideal generated by a d-sequence is of linear type [16].

Finally, we list several examples of ideals that are of linear type.

Example 1.2

1. Complete intersections [43]. Let I be an ideal generated by a regular sequence $f_1, \ldots, f_n$, then

$$\mathcal{R}(I) \cong \mathcal{S}(I) = \frac{R[y_1, \ldots, y_n]}{I_2 \begin{bmatrix} y_1 \cdots y_n \\ f_1 \cdots f_n \end{bmatrix}}.$$

2. Any licci ideal with G_∞, for instance any grade two perfect ideal and grade three perfect Gorenstein ideal with G_∞.
3. Some determinantal ideals:

    a. The ideal of maximal minors of an $n \times (n + 1)$ generic matrix.
    b. The ideal of sub-maximal minors of an $n \times n$ generic matrix [22]. This example is interesting because the ideals in this class do not satisfy the sliding depth condition, nor can they be generated by a d-sequence if $n \ge 3$.
    c. The ideal of sub-maximal minors of an $n \times n$ generic symmetric matrix [26].

4. The defining ideal of the twisted cubic in $\mathbb{P}^3$.
5. The defining ideal of n general points in $\mathbb{P}^2$ with $n \le 5$.
6. Let M be any finite R-module, then $[\text{Sym}(M)]_{>0}$ is of linear type. Indeed, by Valla [51] an ideal I of a Noetherian ring S is of linear type if and only if $G(I) \cong \text{Sym}_{S/I}(I/I^2)$. We can apply this characterization to $I = [\text{Sym}(M)]_{>0} \subset S = \text{Sym}(M)$.
7. Let $\varphi = (x_{ij})$ be an $m \times n$ generic matrix and $R = k[x_{ij}]$ be the polynomial ring in the variables x_{ij} over a field k. Let I_1 and I_2 be ideals of maximal minors of given submatrices of φ. Let

$$\mathbb{D} = \ker(S = R/I_1 \otimes_k R/I_2 \longrightarrow R/(I_1 + I_2)) \subset S.$$

Then the ideal $\mathbb{D}$ is of linear type [38]. Note that the special fiber ring $\mathcal{F}(\mathbb{D})$ is the homogeneous coordinate ring of the embedded join of the varieties $V(I_1)$ and $V(I_2)$ in the projective space $\mathbb{P}^{mn-1}$. Since $\mathbb{D}$ is of linear type, $\mathcal{F}(\mathbb{D})$ is a polynomial ring, thus the embedded join is the whole space. If $I_1 = I_2$, then $\mathcal{F}(\mathbb{D})$ is the coordinate ring of the secant variety.

1.3   Perfect Ideals of Height Two

In Sect. 1.2 we have proven that perfect ideals of grade two satisfying G_∞ are of linear type. In this section, we want to consider perfect ideals of grade two that do not satisfy G_∞, hence cannot be of linear type. However, we assume that they are of linear type on the punctured spectrum, that is, they satisfy a weaker condition called G_d, where d is the dimension of the ring. For instance, let I be the defining ideal of n general points in $\mathbb{P}^2$. If $n = 6$, then I has too many generators to be G_∞ and hence to be of linear type. However, I is a perfect ideal of grade two and is generically a complete intersection, thus I is of linear type on the punctured spectrum of $R = k[x_0, x_1, x_2]$. What is the defining ideal of its Rees algebra?

Definition 1.6 Let s be a non-negative integer. We say that an ideal I in a Noetherian ring R satisfies the property G_s if $\mu(I_\mathfrak{p}) \leq \dim R_\mathfrak{p}$ for all $\mathfrak{p} \in V(I)$ with $\dim R_\mathfrak{p} < s$.

1.3.1   The Jacobian Dual Method

Let $(R, \mathfrak{m}, k)$ be a Gorenstein local (or $*$-local) ring of dimension d. Assume that the residue field k is infinite. Let I be a perfect ideal of height two minimally generated by $f_1, \ldots, f_n$ that satisfies G_d. We may assume that $n > d$, otherwise I satisfies G_∞ and hence it is of linear type. Let φ be a minimal syzygy matrix of $f_1, \ldots, f_n$, that is

$$0 \longrightarrow R^{n-1} \xrightarrow{\ \varphi\ } R^n \xrightarrow{\ [f_1 \ \cdots \ f_n]\ } I \longrightarrow 0 \,.$$

In [50, 5.4], Ulrich gives a complete characterization of when the Rees algebra of I is Cohen-Macaulay in terms of the presentation matrix φ. In turn, the Cohen-Macaulayness of the Rees algebra allows one to control the number and the degrees of the defining equations.

Theorem 1.4 (Ulrich) *Let $(R, \mathfrak{m})$ be a Gorenstein local ring of dimension d with infinite residue field, I be a perfect ideal of height two satisfying G_d, and φ an $n \times n - 1$ matrix minimally presenting I. The Rees algebra $\mathcal{R}(I)$ is Cohen-Macaulay if and only if after elementary row operations there exists an $(n - d) \times (n - 1)$ submatrix φ' of φ such that $I_{n-d}(\varphi) = I_{n-d}(\varphi')$.*

With the stronger hypothesis $I_1(\varphi)^{n-d} = I_{n-d}(\varphi')$, in [44, 1.2], Morey and Ulrich compute explicitly the defining equations of the Rees ring in terms of the Jacobian dual of φ. For any matrix φ with entries in R, a *Jacobian dual* of φ is defined to be a matrix $B(\varphi)$ with linear entries in the polynomial ring $R[y_1, \ldots, y_n]$ such that

$$[y_1, \ldots, y_n] \cdot \varphi = [a_1, \ldots, a_r] \cdot B(\varphi) \,,$$

where $I_1(\varphi) \subset (a_1, \ldots, a_r)$. We will describe their result in the special case of linearly presented ideals.

Let $I \subset R = k[x_1, \ldots, x_d]$ be a perfect ideal of grade two satisfying G_d and assume that I is minimally presented by a linear matrix φ. After elementary row operations there exists an $(n - d) \times (n - 1)$ submatrix φ' of φ such that $I_{n-d}(\varphi') = (x_1, \ldots, x_d)^{n-d}$. Therefore $I_{n-d}(\varphi') = I_{n-d}(\varphi)$, thus the Rees algebra is Cohen-Macaulay. Furthermore $I_1(\varphi) = (x_1, \ldots, x_d)$, hence the stronger assumption $I_{n-d}(\varphi') = I_1(\varphi)^{n-d}$ is satisfied [44, 1.3].

Since $I_1(\varphi) = (x_1, \ldots, x_d)$, a Jacobian dual of φ is defined as a $d \times n - 1$ matrix with linear entries in $R[y_1, \ldots, y_n]$, such that

$$[y_1, \ldots, y_n] \cdot \varphi = [x_1, \ldots, x_d] \cdot B(\varphi).$$

Furthermore, because φ is a matrix of linear forms, the entries of $B(\varphi)$ can be taken from the ring $k[y_1, \ldots, y_n]$; this $B(\varphi)$ is uniquely determined by φ. As the entries of the row vector $[y_1, \ldots, y_n] \cdot \varphi$ generate $\mathcal{L}$, the defining ideal of the symmetric algebra, the entries of the row vector $[x_1, \ldots, x_d] \cdot B(\varphi)$ are contained in $\mathcal{J}$. Cramer's rule shows that $I_d(B(\varphi))$ annihilates the $\mathcal{R}(I)$-ideal generated by $\underline{x} = x_1, \ldots, x_d$, and hence must be contained in $\mathcal{J}$ since $\mathrm{grade}(\underline{x})\mathcal{R}(I) > 0$. Therefore

$$\mathcal{L} + I_d(B(\varphi)) = (\underline{x} \cdot B(\varphi) + I_d(B(\varphi))) \subset \mathcal{J},$$

and if equality holds, then $\mathcal{J}$ is said to have the *expected form* (in the sense of [52, 3.1]). The expected form is the next best possibility for the defining ideal of the Rees ring if I is not of linear type.

Theorem 1.5 (Morey-Ulrich) *Let $(R, \mathfrak{m})$ be a Gorenstein local ring of dimension d with infinite residue field, I be a perfect ideal of height two satisfying G_d, and φ an $n \times (n-1)$ matrix minimally presenting I. Then $\mathcal{R}(I)$ has the expected form if and only if after elementary row operations there exists an $(n - d) \times (n - 1)$ submatrix φ' of φ such that $I_{n-d}(\varphi') = I_1(\varphi)^{n-d}$. In particular, if I is linearly presented then $\mathcal{R}(I)$ has the expected form.*

The main tool used in the proof is residual intersection theory. Under the above assumptions one proves that the ideal $(\underline{x} \cdot B(\varphi) + I_d(B(\varphi)))$ is a geometric residual intersection of $(\underline{x})$, which says in particular that

$$(\underline{x} \cdot B(\varphi) + I_d(B(\varphi))) = (\underline{x} \cdot B(\varphi)) : (\underline{x}).$$

Example 1.3 If I is the defining ideal of $n = \binom{e}{2}$ general points in $\mathbb{P}^2$, then $\mathcal{R}(I)$ is Cohen-Macaulay of the expected form. Indeed, the ideal I is a perfect ideal of height two, is generically a complete intersection, and is linearly presented.

There have been several generalizations of the Morey-Ulrich theorem, see for instance [9, 36, 37]. In [36], the class of ideals studied does not satisfy G_d, but consists of linearly presented ideals with $d + 1$ generators. In [9, 37], the focus is on Rees algebras of linearly presented perfect ideals of height two in three variables.

Even though the statement of Theorem 1.5 is beautiful, the assumption on φ is quite restrictive. Next we survey two more methods to compute the defining equations of the Rees ring for perfect ideal of height two, *the method of divisors* and *Jouanalou duality*. From now on we will always assume that the ideal I is generated by forms of the same degree. We start by recalling the geometric importance of the Rees algebra and the special fiber ring of I in this situation.

1.3.2 Graphs and Images of Rational Maps

In this subsection we do not assume that I has height two. Let $R = k[x_1, \ldots, x_d]$ be a polynomial ring over a field k. Let I be an ideal generated by forms $f_1, \ldots, f_n$ of degree δ. These forms define a rational map Φ between projective spaces with base locus $V(I)$:

$$\Phi: \mathbb{P}_k^{d-1} \xdashrightarrow{\ [f_1 : \ldots : f_n]\ } \mathbb{P}_k^{n-1}$$

where $(x_1 : \ldots : x_d)$ is sent to $(f_1(x_1, \ldots, x_d) : \ldots : f_n(x_1, \ldots, x_d))$. Let X denote the closed image of Φ, that is, the subvariety of $\mathbb{P}^{n-1}$ parametrized by $f_1, \ldots, f_n$. An important problem in commutative algebra, algebraic geometry, and elimination theory is to determine the implicit equations defining $X \subset \mathbb{P}^{n-1}$. The same problem for small n and d has significant applications in computer-aided geometric design and modeling. But to investigate the image of a map it is often advantageous to consider the graph of the map as it reveals more than the image itself. Hence a broader question is to find the implicit equations of the graph of Φ as a subvariety of $\mathbb{P}^{d-1} \times \mathbb{P}^{n-1}$. The bi-homogeneous coordinate ring of the graph is the Rees algebra $\mathcal{R}(I) = R[f_1 t, \ldots, f_n t]$ of the ideal I, whereas the homogeneous coordinate ring $A(X)$ of X is the subalgebra $k[f_1 t, \ldots, f_n t]$, which is isomorphic to the special fiber ring $\mathcal{F}(I) = \mathcal{R}(I) \otimes_R k$ of I. The natural diagram

$$
\begin{array}{ccc}
\text{graph } \Phi & \subset & \mathbb{P}_k^{d-1} \times \mathbb{P}_k^{n-1} \\
\big\downarrow & & \big\downarrow \\
X = \text{im } \Phi & \subset & \mathbb{P}_k^{n-1}
\end{array}
$$

corresponds to the diagram

where S is the bi-homogeneous coordinate ring of the product $\mathbb{P}^{d-1} \times \mathbb{P}^{n-1}$ and T is the homogenous coordinate ring of $\mathbb{P}^{n-1}$. Hence the task becomes to determine the defining ideals $\mathcal{J}$ and $\mathcal{K}$ of $\mathcal{R}(I)$ and $\mathcal{F}(I)$. The Rees algebra inherits a standard bi-grading as a bi-graded coordinate ring, given by $\deg x_i = (1, 0)$ and $\deg f_j t = (0, 1)$. Hence the kernel $\mathcal{J}$ of the epimorphism $S \twoheadrightarrow \mathcal{R}(I)$ is also a bi-homogenous ideal and $I(X)$ is simply $\mathcal{J}_{(0,\star)}$.

1.3.3   The Method of Divisors

In this subsection we treat height two ideals in the polynomial ring $R = k[x_1, x_2]$ that are generated by forms of the same degree δ. By the Hilbert-Burch theorem such an ideal I is generated by the maximal minors of an $n \times (n-1)$ matrix φ with homogenous entries:

$$0 \longrightarrow R^{n-1} \xrightarrow{\ \varphi\ } R^n \xrightarrow{\ [f_1\ \cdots\ f_n]\ } I \longrightarrow 0 .$$

We assume that the columns of φ have degrees $1, \ldots, 1, t$. If $t = 1$, then the ideal I is a power of the homogenous maximal ideal (x_1, x_2) and its Rees ring is described by Theorem 1.5. Hence we assume that $t > 1$. In [18] and [6] this problem was solved for $n = 3$ and $t \leq 5$ and for $n = 3$ and arbitrary t, respectively. The approach utilized in these papers is an iterated use of Jacobian duals or Sylvester forms.

In [29], instead of employing the Jacobian dual method, we produce an ideal H of $S = R[y_1, \ldots, y_n]$ such that $A = S/H$ is a normal domain mapping onto $\mathcal{R}(I)$ of dimension $1 + \dim \mathcal{R}(I)$. This strategy replaces $\mathrm{Sym}(I)$ with the normal domain A as the starting point of our investigation.

The ideals under consideration naturally fall into two classes, depending on ρ, the number of blocks in the Kronecker-Weierstraß normal form of the linear part of φ. After row and column operations the linear part of the matrix φ is equal to the $n \times (n-2)$ matrix

$$\varphi' = \begin{cases} \begin{bmatrix} D_{\sigma_1} & 0 \\ 0 & D_{\sigma_2} \end{bmatrix}, & \text{if } \rho = 2, \\[6mm] \begin{bmatrix} D_{\sigma_1} \\ 0 \end{bmatrix}, & \text{if } \rho = 1, \end{cases}$$

where D_a is the $(a+1) \times a$ matrix

$$D_a = \begin{bmatrix} x_1 & 0 & 0 & \cdots & 0 \\ -x_2 & x_1 & 0 & \cdots & 0 \\ 0 & -x_2 & x_1 & \cdots & 0 \\ \vdots & & \ddots & \ddots & \\ 0 & 0 & \cdots & -x_2 & x_1 \\ 0 & 0 & \cdots & 0 & -x_2 \end{bmatrix},$$

and

$$\begin{cases} \sigma_1 \geq \sigma_2 \geq 1, \text{ and } \sigma_1 + \sigma_2 = n-2, & \text{if } \rho = 2, \text{ or} \\ \sigma_1 = n-2, & \text{if } \rho = 1. \end{cases}$$

This reduction works because we are in a polynomial ring in two variables. The geometric significance of ρ is as following: $\rho = 1$ if and only if φ' has a generalized row of zeros if and only if the variety $X \subset \mathbb{P}^{n-1}$ parametrized by $f_1, \ldots, f_n$ has a singularity of multiplicity t [7]. Whereas $\rho = 2$ if and only if φ' does not have a generalized row of zeros if and only if the variety $X \subset \mathbb{P}^{n-1}$ is smooth [7].

We can think of I as a truncation of a complete intersection generated by a homogeneous regular sequence h_1, h_2 in R:

$$I = ((x_1, x_2)^{\sigma_1} h_1, (x_1, x_2)^{\sigma_2} h_2)$$

and the only non-linear column in φ comes from the syzygy between h_1 and h_2.

The ideal defining the symmetric algebra can be written as

$$\mathcal{L} = (\underline{y} \cdot \varphi) = (\underline{y} \cdot \varphi', g),$$

where $g \in R[y_1, \ldots, y_n]$ is the bi-homogeneous polynomial of degree $(t, 1)$ obtained by multiplying the row vector $\underline{y}$ with the only non-linear column of φ. We consider a partial Jacobian dual $B(\overline{\varphi'})$ constructed using the linear part of φ. The $2 \times (n-2)$ matrix $B(\varphi')$ with linear entries in $k[y_1, \ldots, y_n]$ satisfies

$$\underline{y} \cdot \varphi' = \underline{x} \cdot B(\varphi').$$

Thus, again by Cramer's rule, $I_2(B(\varphi')) \subset \mathcal{J}$. The ideal $(\underline{y} \cdot \varphi', I_2(B(\varphi'))) \subset \mathcal{J}$ can be written as the ideal of 2×2 minors of a generic scroll matrix ψ with two rows and two or three blocks depending on ρ,

$$\psi = \begin{cases} \left[\psi_{\sigma_1} \middle| \psi_{\sigma_2} \middle| \psi_3 \right], & \text{if } \rho = 2, \\ \left[\psi_{\sigma_1} \middle| \psi_3 \right], & \text{if } \rho = 1, \end{cases}$$

where $\psi_3 = \begin{bmatrix} x_2 \\ x_1 \end{bmatrix}$,

$$\psi_{\sigma_1} = \begin{bmatrix} y_1 & y_2 & \cdots & y_{\sigma_1-1} & y_{\sigma_1} \\ y_2 & y_3 & \cdots & y_{\sigma_1} & y_{\sigma_1+1} \end{bmatrix},$$

and if $\rho = 2$,

$$\psi_{\sigma_2} = \begin{bmatrix} y_{\sigma_1+2} & y_{\sigma_1+3} & \cdots & y_{n-2} & y_{n-1} \\ y_{\sigma_1+3} & y_{\sigma_1+4} & \cdots & y_{n-1} & y_n \end{bmatrix}.$$

We write $H = I_2(\psi)$. Thus,

$$(H, g) \subset \mathcal{J}.$$

Let $A = S/H$. While in the previous methods we used the symmetric algebra to approximate the Rees algebra, now we use the normal domain A:

$$0 \longrightarrow \mathcal{A} = \mathcal{J}/H \longrightarrow A \longrightarrow \mathcal{R}(I) \longrightarrow 0.$$

The Krull dimension of A is $4 = \dim \mathcal{R}(I) + 1$, thus the kernel $\mathcal{A}$ is an ideal of height one, necessarily prime since the Rees ring is a domain, hence a divisorial ideal. Our goal then becomes to understand the divisor class group of A. As A is the coordinate ring of a rational normal scroll, the issue is to investigate divisors on rational normal scrolls. This problem, interesting in its own right, is treated in complete generality in [28] (see also [46]).

The divisor class group of A is cyclic, and infinite since ψ has at least two blocks. One generator of $\text{Cl}(A)$ is given by the class of the ideal L generated by the entries of the first column of ψ. The positive powers of L are well understood. Indeed, the powers, the symbolic powers, and the symmetric powers all coincide. The inverse of $[L]$ in $\text{Cl}(A)$ is given by $[K]$, where K is the ideal generated by the entries of the first row of ψ. The Rees ring of I corresponds to a divisor on A. Since its defining ideal $\mathcal{A}$ in A is not isomorphic to a power of L, it must be isomorphic to $K^{(s)}$ for some s. The integer s turns out to be t, the degree of the last column in the presentation matrix φ of I.

Theorem 1.6 (Kustin-P-Ulrich) *In the quotient field of A we have the equality:*

$$\mathcal{A} = \frac{g}{x_2^t} \cdot K^{(t)}.$$

As the isomorphism $\mathcal{A} \cong K^{(t)}$ is explicit, we can exhibit the defining equations on the nose once the generators of $K^{(t)}$ are understood. To determine the generators of $K^{(t)}$ we give the variables different degrees and we identify the t-th symbolic powers of K with $A_{\geq t}$. The grading on S is defined by $\text{Deg}(x_2) = 1$, $\text{Deg}(x_1) = 0$,

$\mathrm{Deg}(y_j) = \sigma_1 + 1 - j$ for $1 \leq j \leq \sigma_1 + 1$, $\mathrm{Deg}(y_j) = n - j$ for $\sigma_1 + 2 \leq j \leq n$. With respect to this grading H is a homogeneous ideal and thus Deg induces a grading on A. Let $A_{\geq t}$ be the ideal of A generated by all monomials m with $\mathrm{Deg}(m) \geq t$:

Theorem 1.7 (Kustin-P-Ulrich) $K^{(t)} = A_{\geq t}$.

From the model $K^{(t)}$ we can deduce the algebraic properties of $\mathcal{R}(I)$. The ideal $K^{(t)}$ is generated by monomials, and hence considerably easier to analyze. Thus we are able to study Serre's properties of the blowup algebras, Hilbert functions, reduction numbers, and the Castelnuovo-Mumford regularity of the powers of I. In [8, 25, 49] the regularity of I^s is shown to be a linear function of s for all $s \gg 0$, i.e., $\mathrm{reg}(I^s) = sd + e$. By a theorem of Eisenbud and Harris the value of e is known and well-understood [11]. The problem remains of when the asymptotic value is obtained, i.e., for which s_0 we have $\mathrm{reg}(I^s) = sd + e$ whenever $s \geq s_0$. If $\rho = 2$, then $e = 0$, the powers of I eventually become powers of the maximal ideal, and $s_0 = \left\lceil \frac{n-1}{\sigma_2} \right\rceil + 1$. The other class, when $\rho = 1$, is very special, for instance the special fiber ring is Cohen-Macaulay. In this case $e = n - 1$ and $s_0 = 1$.

Finally, we introduce a filtration of $K^{(t)}$ whose factors are Cohen-Macaulay modules resolved by Eagon-Northcott complexes, and hence we can resolve $K^{(t)}$. Using this resolution we give an explicit resolution of $\mathcal{R}(I)$ and therefore of all powers of I.

These theorems and the method of divisors have been extended and generalized by [2, 39, 41]. In [2], the class of ideals treated are height two perfect ideals in $R = k[x_1, \ldots, x_d]$ that satisify G_d and have an almost linear presentation matrix. In [39], we study the Rees algebra of truncations of complete intersections. In [41], the dimension of the ring is still two but the presentation matrix φ of I is not assumed to be almost linear, however the results obtained in [41] are not as complete as above.

Problem 1.1 Let k be an algebraically closed field and let X be a general set of e points in $\mathbb{P}^2_k$ [10, Problem 13, page 50]. Write $R = k[x_0, x_1, x_2]$ for the coordinate ring of $\mathbb{P}^2_k$, $I = I_X \subset R$ for the homogeneous ideal of X and $R_X = R/I_X$ for the homogeneous coordinate ring of X. The ideal I is perfect of height two. Furthermore, I satisfies $G_d = G_3$ since I is reduced and $I_\mathfrak{p} = \mathfrak{p}R_\mathfrak{p}$ locally at every prime $\mathfrak{p} \in \mathrm{Min}(I)$. What are the defining equations of $\mathcal{R}(I)$?

There are unique non-negative integers s and t such that $e = \binom{s+t+1}{2} + s$. The problem naturally divides into two cases [10, Problem 13, page 50]. If $s \leq t$, then φ has $t - s$ linear columns and s quadratic columns. If $t \leq s$, then φ has $s - t$ linear rows and $t + 1$ quadratic rows . Notice if $\max\{t, s\} \leq 2$ then the ideal I is of linear type. In general, if $t \leq s$, we may use the method of divisors as in [2] (at least when $s = 1$) and if $s \leq t$, we may use the method of Jacobian duals as in [44] (at least when t is small). In collaboration with Eisenbud and Ulrich we have worked out this problem for $s \leq 1$ and $t \leq 2 < s$.

1.3.4 Jouanalou Duality and Rational Plane Curves

This method works if the symmetric algebra of I is a complete intersection, that is for perfect ideals of grade two with $\mu(I_{\mathfrak{p}}) \leq \dim(R_{\mathfrak{p}}) + 1$. We will focus on the case of three generated ideals in a polynomial ring in two variables over an algebraically closed field k. Moreover we assume that f_1, f_2, f_3, the generators of I, are forms of the same degree δ. Dividing by a greatest common divisor we may harmlessly assume that f_1, f_2, f_3 generate an ideal I of height two. This is the setting of rational plane curves. Indeed the map $\Phi = [f_1 : f_2 : f_3]$ parametrizes a rational plane curve $C \subset \mathbb{P}_k^2$ that we can assume has degree δ [30], and $\mathcal{R}(I)$ is the bi-homogeneous coordinate ring of the graph of Φ. The problem of describing the equations defining $\mathcal{R}(I)$ is very difficult, even in this simple situation. There is a correspondence between local properties of C (such as the configuration of its singularities) and the bidegrees of the polynomials defining $\mathcal{R}(I)$ [7, 32].

The syzygy matrix φ of f_1, f_2, f_3 gives a homogeneous free resolution of the ideal I,

$$0 \longrightarrow R(-d_1) \oplus R(-d_2) \xrightarrow{\varphi} R^3 \longrightarrow I(\delta) \longrightarrow 0 .$$

The matrix φ is a 3×2 matrix with homogeneous entries in R, of degree d_1 in the first column and of degree d_2 in the second column. Notice that $\delta = d_1 + d_2$ according to the Hilbert-Burch Theorem. We may assume that $d_1 \leq d_2$. If $d_1 = 1$, the ideal $\mathcal{J}$ defining the Rees algebra can be described explicitly using the method of divisors. Thus we may assume that $d_1 \geq 2$.

As usual we study the Rees algebra via the symmetric algebra,

$$0 \longrightarrow \mathcal{A} \longrightarrow \mathcal{S}(I) \longrightarrow \mathcal{R}(I) \longrightarrow 0 .$$

Since R/I is zero-dimensional, the Rees algebra and the symmetric algebra of I are isomorphic locally at every prime $\mathfrak{p} \in \mathrm{Spec}(R)$ with $\mathfrak{p} \neq \mathfrak{m} = (x_1, x_2)$. Thus

$$\mathcal{A} = H_{\mathfrak{m}}^0(\mathcal{S}(I)) = 0 :_{\mathcal{S}(I)} \mathfrak{m}^\infty .$$

As the ideal I is three generated, its symmetric algebra $\mathcal{S}(I)$ is a complete intersection. More precisely, letting $S = R[y_1, y_2, y_3]$ be a standard bigraded polynomial ring over k with $\deg x_i = (1, 0)$ and $\deg y_i = (0, 1)$, and mapping the variables y_i to homogeneous minimal generators of I, one obtains a presentation $\mathcal{S}(I) \cong S/(\ell_1, \ell_2)$, where ℓ_1, ℓ_2 is an S-regular sequence of forms of bidegrees $(d_1, 1), (d_2, 1)$. This isomorphism endows $\mathcal{S}(I)$ with the standard bigrading over k. The Koszul complex of ℓ_1, ℓ_2 provides a bi-homogeneous free S-resolution of $\mathcal{S}(I)$:

$$K_\bullet(\ell_1, \ell_2; S) \longrightarrow \mathcal{S}(I) \longrightarrow 0.$$

Computing local cohomology with support in $\mathfrak{m}$ along this resolution, using the symmetry of the Koszul complex, and the isomorphism $H_{\mathfrak{m}}^2(R) \cong \underline{\mathrm{Hom}}_k(R, k)(2)$, where <u>Hom</u> denotes the graded dual, we obtain an isomorphism between $\mathcal{A} = H_m^0(S(I))$ and the graded dual $\underline{\mathrm{Hom}}_T(S(I), T)(2 - \delta, -2)$ of $S(I)$ as a module over $T = k[y_1, y_2, y_3]$:

Proposition 1.2 *One has* $\mathcal{A} = H_m^0(S(I)) \cong \underline{\mathrm{Hom}}_T(S(I), T)(2 - \delta, -2)$.

This isomorphism is made explicit in [24] using *Morley forms*. Notice that the ideal $\mathcal{A}$ is concentrated in finitely many x-degrees because it is annihilated by a power of m.

The goal is to compute the graded components of $\mathcal{A}$ with respect to the x-grading, $\mathcal{A}_{(i,\star)} = \bigoplus_j \mathcal{A}_{i,j}$. Recall that $\mathcal{A}_{(0,\star)}$ is the defining ideal of the coordinate ring $A(C) \cong \mathcal{F}(I)$. This ideal is generated by the resultant $\mathrm{Res}_{\{x_1,x_2\}}(\ell_1, \ell_2)$. Thus we may assume that $i > 0$. Proposition 1.2 shows that there is an isomorphism of graded T-modules

$$\mathcal{A}_{(i,\,\star)} \cong \mathrm{Hom}_T(S(I)_{(\delta-2-i,\,\star)}, T)(-2) . \tag{1.3}$$

From this isomorphism we deduce that the T-modules $\mathcal{A}_{(i,\,\star)}$ are zero for all $i > \delta - 2$, are reflexive for all i, and their module structure is completely determined by the one of $S(I)_{(\delta-2-i,\,\star)}$. Since the symmetric algebra is a complete intersection defined by the regular sequence ℓ_1, ℓ_2 of bidegree $(d_1, 1)$ and $(d_2, 1)$, the T-module structure of $S(I)_{(\delta-2-i,\,\star)}$ depends on the relationship between $\delta - 2 - i$, d_1, and d_2. We have three ranges to consider. We will describe them below.

Write the polynomials ℓ_1 and ℓ_2 in $T[x_1, x_2]$ as

$$\ell_i = \sum_{j=0}^{d_i} c_{j,i} x_1^j x_2^{d_i - j}$$

with $c_{j,i}$ linear polynomials in $T = k[y_1, y_2, y_3]$. For positive integers n and i, with i equal to 1 or 2, let $C_{n,i}$ be the $(n + d_i) \times n$ matrix

$$C_{n,i} = \begin{bmatrix} c_{0,i} & 0 & 0 & \cdots & 0 \\ c_{1,i} & c_{0,i} & 0 & \cdots & 0 \\ c_{2,i} & c_{1,i} & c_{0,i} & \cdots & 0 \\ \vdots & & \ddots & \ddots & \vdots \\ & & & & c_{0,i} \\ \vdots & & & & \vdots \\ c_{d_i,i} & & & & \\ 0 & c_{d_i,i} & & & \\ 0 & 0 & c_{d_i,i} & & \\ \vdots & & & \ddots & \vdots \\ 0 & 0 & 0 & \cdots & c_{d_i,i} \end{bmatrix},$$

with linear entries from $T = k[y_1, y_2, y_3]$. The matrix $C_{n,i}$ represents the map of free T-modules

$$T[x_1, x_2]_{n-1} \to T[x_1, x_2]_{n-1+d_i}$$

that is given by multiplication by ℓ_i when the bases

$$x_2^{n-1}, \ldots, x_1^{n-1} \quad \text{and} \quad x_2^{n-1+d_i}, \ldots, x_1^{n-1+d_i}$$

are used for $T[x_1, x_2]_{n-1}$ and $T[x_1, x_2]_{n-1+d_i}$, respectively.

We can describe the graded T-module $\mathcal{A}_{(i, \star)}$ in terms of $S(I)_{(\delta-2-i, \star)}$ in each range of i as follows:

1. If $0 \leq i \leq d_1 - 2$, then the T-modules $\mathcal{A}_{(i, \star)}$ and $S(I)_{(\delta-2-i, \star)}$ both have rank $i + 1$. Furthermore the matrix $\left(C_{d_2-i-1,1} \; C_{d_1-i-1,2}\right)$ gives a T-resolution of $S(I)_{(\delta-2-i, \star)}$ and the following sequence is exact:

$$0 \to \mathcal{A}_{(i, \star)} \to T(-2)^{\delta-2-i+1} \xrightarrow{\begin{pmatrix} C_{d_2-i-1,1}{}^{\mathrm{T}} \\ C_{d_1-i-1,2}{}^{\mathrm{T}} \end{pmatrix}} T(-1)^{d_2-i-1} \oplus T(-1)^{d_1-i-1}.$$

2. If $d_1 - 1 \leq i \leq d_2 - 2$, then the T-modules $\mathcal{A}_{(i, \star)}$ and $S(I)_{(\delta-2-i, \star)}$ both have rank d_1. Furthermore the matrix $C_{d_2-i-1,1}$ gives a T-resolution of $S(I)_{(\delta-2-i, \star)}$ and the following sequence is exact:

$$0 \to \mathcal{A}_{(i, \star)} \to T(-2)^{\delta-2-i+1} \xrightarrow{C_{d_2-i,1}{}^{\mathrm{T}}} T(-1)^{d_2-i-1}.$$

3. If $d_2 - 1 \leq i \leq \delta - 1$, then the T-modules $\mathcal{A}_{(i, \star)}$ and $S(I)_{(\delta-2-i, \star)}$ both have rank $\delta - i - 1$. Furthermore,

$$S(I)_{(\delta-2-i, \star)} \cong T^{\delta-i-1} \quad \text{and} \quad \mathcal{A}_{(i, \star)} \cong T(-2)^{\delta-i-1}.$$

In the third range, the $S(I)$-ideal $\mathcal{A}_{(\geq d_2-1, \star)}$ is generated in degree $(d_2 - 1, 2)$, in particular the minimal generators of $\mathcal{A}$ all have x-degree at most $d_2 - 1$. The generators of $\mathcal{A}_{(\geq d_2-1, \star)}$ can be easily computed using linkage; indeed

$$\mathcal{A}_{(\geq d_2-1, \star)} = 0 :_{S(I)} \mathfrak{m}^{d_1} = \frac{(\ell_1, \ell_2)S :_S \mathfrak{m}^{d_1}}{(\ell_1, \ell_2)S}$$

where $(\ell_1, \ell_2)S :_S \mathfrak{m}^{d_1}$ is a link of a perfect ideal of grade 2, because ℓ_1, ℓ_2 is a regular sequence contained in $\mathfrak{m}^{d_1} S$.

The first and the second range are much more interesting. In [32], we show that if $d_1 - 1 \leq i \leq d_2 - 2$, then $\mathcal{A}_{(i, \star)}$ is free of rank d_1 if and only if the first column of the matrix φ has a generalized zero if and only if the curve C has a singularity of multiplicity d_2. In this case $\mathcal{A}_{(i,\star)}$ is generated in at most two degrees, and we can describe the bidegrees of the generators of $\mathcal{A}_{(\geq d_1-1,\star)}$ as a T-module and as an S-module.

The first range, namely $1 \leq i \leq d_1 - 2$, is the hardest to handle because the structure of $S(I)_{(\delta-2-i,\star)}$ is more complicated as it depends on both forms ℓ_1 and ℓ_2. If $d_1 = d_2$, the T-module $S(I)_{(d_1,\star)}$ is presented by C, a $2 \times (d_1 + 1)$ matrix with linear entries in T. This matrix is the key instrument to distinguish between the seven possible configurations of multiplicity $c = d_1$ singularities on or infinitely near the curve C. Again, to determine $\mathcal{A}_{(d_1-2,\star)}$ it suffices to compute the kernel of the transpose of C. To compute this kernel we classify all possible $2 \times (d_1 + 1)$ matrices with linear entries in y_1, y_2, y_3. This classification only depends on the minimal number of generators of $I_2(C)$ – the same invariant that detects the number of multiplicity c singularities on or infinitely near the curve C [7]. There are at least two such singularities if and only if $\mathcal{A}_{(d_1-2,\star)}$ is free [32].

The results outlined in this section suffice to provide explicit defining equations for $\mathcal{R}(I)$ if $\delta = d_1 + d_2 \leq 6$, since then $d_1 \leq 2$ or $d_1 = d_2$. We focus on the case $\delta = 6$, the case of a sextic curve.

As it turn out, there is, essentially, a one-to-one correspondence between the bidegrees of the defining equations of $\mathcal{R}(I)$ on the one hand and the types of singularities on or infinitely near the curve C on the other hand. Here one says that a singularity is infinitely near C if it is obtained from a singularity on C by a sequence of quadratic transformations [7]. The correspondence is summarized in the following chart; the first column gives the possible values of d_1, d_2, namely 1, 5 or 2, 4 or 3, 3; the second column lists the corresponding bidegrees of minimal generators of the defining ideal $\mathcal{J}$ together with the multiplicities by which they appear, suppressing the obvious bidegrees $(d_1, 1)$, $(d_2, 1)$ (of the equations defining $S(I)$) and $(0, 6)$ (of the implicit equation of C); this second column reflects the results of the present section; the third column, finally, gives the multiplicities of the singularities on or infinitely near C.

$d_1\ d_2$	equations of $\mathcal{R}$	singularities of C
1 5	$(4, 2)\ (3, 3)\ (2, 4)\ (1, 5)$	' 1 of multiplicity 5 on C
2 4	$(2, 2)\ \ 2(1, 3)$	1 of multiplicity 4 on C
		4 double points on or near C
	$(3, 2)\ \ 3(2, 3)\ \ 4(1, 4)$	10 double points on or near C
3 3	$3(2, 2)\ \ 4(1, 4)$	10 double points on or near C
	$3(2, 2)\ \ (1, 3)\ \ 2(1, 4)$	1 of multiplicity 3 on C
		7 double points on or near C
	$3(2, 2)\ \ 2(1, 3)$	2 of multiplicity 3 and
		4 double points on or near C
	$(2, 2)\ \ (1, 2)\ \ (1, 4)$	3 of multiplicity 3 and
		1 double point on or near C

Madsen [40] has computed explicitly the ideal $\mathcal{J}$ when $\delta = 7$ and has established a similar correspondence.

1.4 Bounding Bidegrees of Defining Equations

In this section we consider a more general class of ideals. We cannot hope to describe explicitly the ideal $\mathcal{J}$ in this case, however we want to bound the bidegrees of the equations defining $\mathcal{R}(I)$. Let I be any ideal of $R = k[x_1, \ldots, x_d]$ generated by forms $f_1, \ldots, f_n$ of degree δ. We denote by $X \subset \mathbb{P}^{n-1}$ the variety parametrized by $\Phi = [f_1, \ldots, f_n]$ and by $I(X)$ the defining ideal of X, that is, the defining ideal of $\mathcal{F}(I)$. Write $\mathfrak{m}$ for the homogenous maximal ideal of R.

As in the previous sections, we map the symmetric algebra to the Rees algebra

$$0 \longrightarrow \mathcal{A} = \mathcal{J}/\mathcal{L} \longrightarrow \mathcal{S}(I) \longrightarrow \mathcal{R}(I) \longrightarrow 0 .$$

We assume that I is of linear type on the punctured spectrum. This is equivalent to

$$\mathcal{A} = H^0_{\mathfrak{m}}(\mathcal{S}(I)) = 0 :_{\mathcal{S}(I)} \mathfrak{m}^\infty ,$$

or to the fact that $\mathcal{A}$ is concentrated in only finitely many x-degrees. So we want to establish upper bounds for the x-concentration degree and for the x-generation degree of $\mathcal{A}$, that is, we want to compute the highest x-degree of a non-zero element of $\mathcal{A}$ and the highest x-degree of a minimal bi-homogeneous generator of $\mathcal{A}$.

Bounds on the (concentration or generation) degree of $\mathcal{A}$ can provide structural information and are an important step in describing $\mathcal{A}$ explicitly [34]. Furthermore, degree bounds make a difference computationally. Indeed, if $\mathcal{A}$ is concentrated in x-degrees at most s, then the symmetric algebra and the Rees algebra coincide after that degree. In particular, the graded components of either algebra in x-degree $s + 1$ have the same annihilator in the polynomial ring $T = k[y_1, \ldots, y_n]$. However, we can express $I(X)$ in terms of this annihilator, namely

$$I(X) = \mathrm{ann}_T (\mathcal{R}(I)_{(0,\star)}) = \mathrm{ann}_T (\mathcal{R}(I)_{(s+1,\star)}) .$$

Thus $I(X)$ is identified with the annihilator of a module, $\mathcal{S}(I)_{(s+1,\star)}$, with a known presentation or even a known resolution.

Bounds on the generation degree of $\mathcal{A}$ also allow to determine when I is of *fiber type*. Recall that even though $\mathcal{J}$ always encodes the ideal $I(X)$, as $I(X) = \mathcal{J}_{(0,\star)} = \mathcal{A}_{(0,\star)}$, the ideal $I(X)$ in general does not determine $\mathcal{J}$. If the ideal $\mathcal{J}$ is generated by the equations defining the symmetric algebra and the equations defining the special fiber ring of I, that is $\mathcal{J} = \mathcal{L} + I(X)\mathcal{S}$, then we say that I is of *fiber type*. We can express this condition in terms of $\mathcal{A}$:

Definition 1.7 I is of *fiber type* if $\mathcal{A} = I(X) \cdot S(I)$, or equivalently, if $\mathcal{A}$ is generated in degree $(0, \star)$.

As $\mathcal{A} = H_{\mathfrak{m}}^0(S(I))$, the ideal I is of fiber type if and only if $H_{\mathfrak{m}}^0(S(I))$ is generated in x-degree zero. Or better yet, if and only if $H_{\mathfrak{m}}^0(S_j(I(\delta)))$ is generated in degree zero for every j. Thus in general the problem becomes to bound the (zeroth) local cohomology modules of the symmetric powers $S_j(I(\delta))$. The advantage of considering $S_j(I(\delta))$ is that the latter is finitely generated as an R-module.

Fix j and let $M = S_j(I(\delta))$. If $\mathbb{F}_\bullet$ is a minimal homogenous free R-resolution of M, then the socle of M is isomorphic to $k \otimes F_d(d)$. In particular, the socle degrees of M can be read from the shifts at the end of a minimal homogeneous finite free resolution of M. Of course, the socle of M is the socle of the local cohomology module $H_{\mathfrak{m}}^0(M)$, and the top degree of the socle of $H_{\mathfrak{m}}^0(M)$ is the top degree of $H_{\mathfrak{m}}^0(M)$. Thus we attain

$$\operatorname{topdeg} H_{\mathfrak{m}}^0(M) = b(F_d) - d\,,$$

where $b(F_d)$ is the largest generator degree of F_d. However, in general we do not have a minimal homogenous free R-resolution of M. As we saw in Sect. 1.2, *approximate resolutions* of $M = S_j(I(\delta))$ may be more readily available. Hence we want to find bounds on the concentration degree and generation degree of $H_{\mathfrak{m}}^0(M)$ in terms of an approximate resolution $C_\bullet$ of M, that is, of a graded complex of finite modules with $H_0(C_\bullet) = M$ so that the modules C_j have high depth and $H_j(C_\bullet)$ have small Krull dimension for $j > 0$.

Definition 1.8 An *approximate resolution* of M is a homogeneous complex of finite R-modules

$$C_\bullet: \qquad \cdots \longrightarrow C_1 \longrightarrow C_0 \longrightarrow 0$$

with $H_0(C_\bullet) = M$, $\dim H_j(C_\bullet) \leq j$ for $j > 0$, and $\operatorname{depth} C_j > j$ for $0 \leq j \leq d - 1$.

The first result, which was observed before [5, 14], provides an upper bound for the concentration degree of $H_{\mathfrak{m}}^0(M)$. A proof of this theorem can be given with a spectral sequence argument.

Theorem 1.8 (Gruson-Lazarsfeld-Peskine, Chardin, Kustin-Polini-Ulrich) *If $C_\bullet$ is an approximate resolution of M, then $H_{\mathfrak{m}}^0(M)$ is concentrated in degrees at most $b(C_d) - d$.*

The second result is more interesting. It gives an upper bound for the generation degree of $H_{\mathfrak{m}}^0(M)$ in terms of the highest generator degree of C_{d-1}. This sharper estimate is more difficult to prove and more relevant for applications to the Rees ring [33, 34]. For instance, it can be used to prove that an ideal is of fiber type.

Theorem 1.9 (Kustin-Polini-Ulrich) *If $C_\bullet$ is an approximate resolution of M, then $H^0_{\mathfrak{m}}(M)$ is generated in degrees at most $b(C_{d-1}) - d + 1$.*

How do we produce approximate resolutions of the symmetric powers of arbitrary ideals? We now give several examples where Theorems 1.8 and 1.9 have been applied and we describe the complexes that serve as approximate resolutions.

Example 1.4

1. This example can be found in [34]. Let I be a perfect ideal of height two generated by forms of the same degree δ. Let φ be a homogenous Hilbert-Burch matrix minimally presenting I with column degrees $\varepsilon_1 \geq \varepsilon_2 \geq \cdots \geq \varepsilon_{n-1}$,

$$0 \longrightarrow \oplus_{j=1}^{n-1} R(-\varepsilon_j) \xrightarrow{\ \varphi\ } R^n \longrightarrow I(\delta) \longrightarrow 0.$$

   If I satisfies G_d, then $\mathcal{A} = H^0_{\mathfrak{m}}(S(I))$ is concentrated in x-degrees $\leq \sum_{j=1}^{d} \varepsilon_j - d$ and is generated in x-degrees $\leq \sum_{j=1}^{d-1} \varepsilon_j - d + 1$. The approximate resolutions used are strands of the Koszul complex $\mathbb{K}_\bullet$ of the defining equations of the symmetric algebra $[\ell_1, \ldots, \ell_{n-1}] = [y_1, \ldots, y_n] \cdot \varphi$. The ideal I is perfect of height two and satisfies G_d. It follows that I is of linear type on the punctured spectrum of R (see Sect. 1.2) and therefore the sequence $\ell_1, \ldots, \ell_{n-1}$ is a regular sequence locally on the punctured spectrum of R. Hence the Koszul complex $\mathbb{K}_\bullet$ is acyclic on the punctured spectrum of R.

2. This example can be found in [34] and it is instrumental to describe the defining equations of the Rees algebra of linearly presented perfect Gorenstein ideals of height three satisfying G_d. Let I be a perfect Gorenstein ideal of height three. Assume that every entry of a homogeneous alternating matrix minimally presenting I has degree ε:

$$R^n(-\varepsilon) \longrightarrow R^n \longrightarrow I(\delta) \longrightarrow 0.$$

   If I satisfies G_d, then $\mathcal{A} = H^0_{\mathfrak{m}}(S(I))$ is concentrated in x-degrees

$$\leq \begin{cases} d(\varepsilon - 1) & \text{when } d \text{ is odd} \\ d(\varepsilon - 1) + \frac{n-d+1}{2}\varepsilon & \text{when } d \text{ is even}, \end{cases}$$

   and is generated in x-degrees $\leq (d - 1)(\varepsilon - 1)$. In particular, if I is linearly presented, that is $\varepsilon = 1$, then I is of fiber type. The approximate resolutions of the symmetric powers used here are the complexes of [27].

3. This example is part of a recent ongoing work with Bernd Ulrich. Assume that I is an ideal of dimension at most one which is generically a complete intersection. Then $\mathcal{A} = H^0_{\mathfrak{m}}(S(I))$ is concentrated in x-degrees $\leq d(\operatorname{reg} I - \delta)$ and is generated in x-degrees $\leq (d - 1)(\operatorname{reg} I - \delta)$. In this situation the approximate resolutions of the symmetric powers are obtained from the construction of [54] applied to a minimal free resolution of I.

4. In the setting of the previous example, Chardin [5] shows that $\mathcal{A} = H^0_{\mathfrak{m}}(S(I))$ is concentrated in x-degrees $\leq (d-1)(\delta-1) - 1$. As approximate resolutions of the symmetric powers he uses strands of the $\mathcal{Z}_\bullet$-complex described in Sect. 1.2. Unfortunately, in general the approximation complexes do not satisfy the depth assumption required for an approximate resolution.

A different approach to bound the x-concentration degree and the x-generation degree of $\mathcal{A}$ is based on estimates for the Castelnuovo-Mumford regularity of Tor-modules. This result appeared first in [12] and has been further generalized by Ulrich and myself in work in progress.

Theorem 1.10 *Assume that the resolution of I is linear for the first $\lceil \frac{d}{2} \rceil$ steps.*

1. *(Eisenbud-Huneke-Ulrich) If I has dimension zero, then $\mathcal{A}$ is concentrated in x-degree zero, namely I is of fiber type and $\mathcal{J} = \mathcal{L} :_S \mathfrak{m}$.*
2. *(Polini-Ulrich) If I has dimension at most one, then $\mathcal{A}$ is generated in x-degree zero, namely I is of fiber type.*

There are many examples of ideals of fiber type to which the above techniques do not apply, for instance maximal minors of generic matrices, see for instance [3].

Conjecture 1.1 Let X be an $m \times n$ matrix of variables and let $1 \leq t \leq \min\{m, n\}$. The ideal $I_t(X) \subset k[X]$ of $t \times t$ minors of X is of fiber type.

If I is linearly presented and $\mathcal{J}$ has the expected form, that is, $\mathcal{J} = (x \cdot B(\varphi)) + I_d(B(\varphi))$, then I is of fiber type. The converse is not true. However, if $\mathcal{A}$ is concentrated in x-degree zero, for instance if I has dimension zero and a linear resolution for the first $\lceil \frac{d}{2} \rceil$ steps, we can show that $\mathcal{J}$ is of the expected form up to radical. Often this suffices to give structural information about the defining equations of $\mathcal{R}(I)$ or $\mathcal{F}(I)$.

Theorem 1.11 *If I is linearly presented and $\mathcal{A}$ is concentrated in x-degree zero, then*

$$\mathcal{J} = \sqrt{\mathcal{L} + I_d(B(\varphi))} \qquad and \qquad I(X) = \sqrt{I_d(B(\varphi))}.$$

Proof Write $B = B(\varphi)$. The ideal $\mathcal{L} :_S \mathfrak{m}$ is the annihilator of the S-module $M = \mathfrak{m}S/\mathcal{L}$. Since $\mathcal{A}$ is concentrated in x-degree zero, $\mathcal{J}$ is the ideal $\mathcal{L} :_S \mathfrak{m}$. Hence $\mathcal{J}$ is the annihilator of M. The ideal $\mathfrak{m}S$ is generated by the entries of $\underline{x}$ and the ideal $\mathcal{L}$ is generated by the entries of $\underline{x} \cdot B$. It follows that M is presented by $\left[\Pi \big| B \right]$, where Π is a presentation matrix of $\underline{x}$ with entries in $\mathfrak{m}$. Therefore

$$\mathcal{J} = \mathrm{ann}_S M \subset \sqrt{\mathrm{Fitt}_0(M)} \subset \sqrt{(\mathfrak{m}, I_d(B))}.$$

Intersecting with $T = k[y_1, \ldots, y_n]$ we obtain

$$I(X) = \mathcal{J} \cap T = \mathcal{J}_{(0,\star)} \subset \left(\sqrt{(\mathfrak{m}, I_d(B))} \right)_{(0,\star)} = \sqrt{I_d(B)} \subset I(X),$$

where the last equality holds because I is linearly presented. Now since I is of fiber type, it follows that

$$\mathcal{J} = (\mathcal{L}, I(X)) = \sqrt{\mathcal{L} + I_d(B)},$$

as desired. □

The same result can be attained more generally; it suffices that one symmetric power of I has an approximate resolution that is linear for the first d-steps [34]:

Theorem 1.12 (Kustin-Polini-Ulrich) *Assume that $\mathcal{A} = H^0_\mathfrak{m}(\mathcal{S}(I))$. If $\mathcal{S}_t(I(\delta))$, for some $t \gg 0$, has an approximate free resolution that is linear for the first d-steps, then*

$$\mathcal{J} = \sqrt{\mathcal{L} + I_d(B(\varphi))} \qquad and \qquad I(X) = \sqrt{I_d(B(\varphi))}.$$

Remark 1.1 The previous theorem applies as long as t is at least the relation type of I. The *relation type* of an ideal is the highest y-degree of a minimal homogeneous generator of $\mathcal{J}$. For instance if I is of linear type, then its relation type is one.

1.4.1 Linearly Presented Height Three Gorenstein Ideals with G_d

We end by describing the defining ideal of the Rees algebra of a linearly presented height three Gorenstein ideal satisfying G_d. Let $R = k[x_1, \ldots, x_d]$ be a polynomial ring over a field and let I be a height three Gorenstein ideal with n homogenous minimal generators $f_1, \ldots, f_n$ and a linear presentation matrix φ. Notice that n is necessarily odd and φ can be assumed to be an $n \times n$ alternating matrix (see [4]). In this case, $\mathcal{J}$ does not have the expected form in general. In fact, the alternating property of φ is responsible for 'unexpected' elements in $\mathcal{J}$.

Write $I(X)$ for the ideal defining the variety X parametrized by the forms $f_1, \ldots, f_n$ that generate the ideal I. We may assume that $d < n$, otherwise I is of linear type as we saw in Sect. 1.2. Let $B = B(\varphi) \in \text{Mat}_{d \times n}(k[y_1, \ldots, y_n])$ be the Jacobian dual of φ. Notice that B is unique since φ is linear. Using the matrices φ and B we build an $(n + d) \times (n + d)$ alternating matrix $\mathcal{B}$. Since φ is alternating and $\underline{y} \cdot \varphi = \underline{x} \cdot B$, it follows that

$$\underline{x} \cdot B \cdot \underline{y}^t = \underline{y} \cdot \varphi \cdot \underline{y}^t = 0.$$

However $B \cdot \underline{y}^t$ has no entries in R, thus $B \cdot \underline{y}^t = 0$. Write

$$\mathcal{B} = \begin{bmatrix} \varphi & B^t \\ -B & 0 \end{bmatrix}.$$

Since $\underline{y} \cdot \varphi - \underline{x} \cdot B = 0$ and $\underline{y} \cdot B^t = 0$, we attain

$$[\underline{y}, \underline{x}] \cdot \mathcal{B} = 0.$$

From the latter we deduce that $(y_1, \ldots, y_n)F \subset I_d(B)$ for some submaximal Pfaffian F of $\mathcal{B}$.

Considering F as a polynomial in the variables $x_1, \ldots, x_d$ and writing $C(\varphi) \subset k[y_1, \ldots, y_n]$ for its content ideal, we obtain the inclusion

$$(y_1, \ldots, y_n) \, C(\varphi) \subset I_d(B)$$

in the ring $k[y_1, \ldots, y_n]$. Since $I(X)$ is a prime ideal in this ring and contains $I_d(B)$, it follows that

$$C(\varphi) \subset I(X) \subset \mathcal{J}.$$

Theorem 1.13 *Let I be a linearly presented height three Gorenstein ideal that satisfies G_d. Then the defining ideal of the Rees ring of I is*

$$\mathcal{J} = \mathcal{L} + I_d(B)S + C(\varphi)S$$

and the defining ideal of the variety X is

$$I(X) = I_d(B) + C(\varphi),$$

where φ is a minimal homogeneous alternating presentation matrix for I. In particular, I is of fiber type and, if d is odd, then $C(\varphi) = 0$ and $\mathcal{J}$ has the expected form.

Recall that I is of fiber type as shown in Example 4.2. Thus it suffices to prove that $I(X) = I_d(B) + C(\varphi)$. To show that the containment $I_d(B) + C(\varphi) \subset I(X)$ is an equality, we first establish that the two ideals have the same radical and hence the same codimension, namely $n - d$. The latter follows from Theorem 12. One uses the complexes of [27] to see that a large symmetric power of I has an approximate resolution that is linear for the first d steps.

The next step is to show that the ideal $I_d(B) + C(\varphi)$ is unmixed. In [35] we construct families of complexes associated to any $d \times n$ matrix B with linear entries that is annihilated by a vector of indeterminates, $B \cdot \underline{y}^t = 0$. The ideal of $d \times d$ minors of such a matrix cannot have generic height $n - d + 1$. However, if the height is $n - d$, our complexes give resolutions of $I_d(\varphi)$ and of $I_d(B) + C(\varphi)$, although these ideals fail to be perfect. Using these resolutions we prove that the latter ideal is the unmixed part of the former and that both ideals are perfect locally on the punctured spectrum. We also compute the multiplicity of the rings defined by these ideals. On the other hand, from results in [31] and [30] we can determine the degree of the variety X. We conclude that the ideals $I_d(B) + C(\varphi) \subset I(X)$

are unmixed of the same height and define rings of the same multiplicity. Thus they have to be equal.

Acknowledgments The author "Claudia Polini" was partially supported by NSF grant DMS-1902033.

References

1. R. Apéry, Sur certains caractéres numériques d'un idéal sans composant impropre. C. R. Acad. Sci. Paris **220**, 234–236 (1945)
2. J. Boswell, V. Mukundan, Rees algebras and almost linearly presented ideals. J. Algebra **460**, 102–127 (2016)
3. W. Bruns, A. Conca, M. Varbaro, Maximal minors and linear powers. J. Reine Angew. Math. **702**, 41–53 (2015)
4. D. Buchsbaum, D. Eisenbud, Algebra structures for finite free resolutions, and some structure theorems for ideals of codimension 3. Am. J. Math. **99**, 447–485 (1977)
5. M. Chardin, *Powers of Ideals: Betti Numbers, Cohomology and Regularity.* Commutative Algebra, Expository Papers Dedicated to David Eisenbud on the Occasion of His 65th Birthday. I. Peeva Ed. (Springer, 2013), pp. 317–333
6. D. Cox, J.W. Hoffman, H. Wang, Syzygies and the Rees algebra. J. Pure Appl. Algebra **212**, 1787–1796 (2008)
7. D. Cox, A. Kustin, C. Polini, B. Ulrich, A study of singularities on rational curves via syzygies. Mem. Am. Math. Soc. **222** (2013)
8. S.D. Cutkosky, J. Herzog, N.V. Trung, Asymptotic behaviour of the Castelnuovo-Mumford regularity. Compositio Math. **118**, 243–261 (1999)
9. A. Doria, Z. Ramos, A. Simis, Linearly presented perfect ideals of codimension 2 in three variables. J. Algebra **512**, 216–251 (2018)
10. D. Eisenbud, *The Geometry of Syzygies. A Second Course in Commutative Algebra and Algebraic Geometry.* Graduate Texts in Mathematics, vol. 229 (Springer, New York, 2005), xvi+243 pp.
11. D. Eisenbud, J. Harris, Powers of ideals and fibers of morphisms. Math. Res. Lett. **17**, 267–273 (2010)
12. D. Eisenbud, C. Huneke, B. Ulrich, The regularity of Tor and graded Betti numbers. Am. J. Math. **128**, 573–605 (2006)
13. F. Gaeta, Quelques progrès récents dans la classification des variétés algébriques d'un espace projectif. Deuxieme Collogue de Géométrie Algébrique Liège. C.B.R.M., 145–181 (1952)
14. L. Gruson, R. Lazarsfeld, C. Peskine, On a theorem of Castelnuovo, and the equations defining space curves. Invent. Math. **72**, 491–506 (1983)
15. J. Herzog, A. Simis, W.V. Vasconcelos, Approximation complexes of blowing-up rings. J. Algebra **74**, 466–493 (1982)
16. J. Herzog, A. Simis, W.V. Vasconcelos, Approximation complexes of blowing-up rings. II. J. Algebra **82**, 53–83 (1983)
17. J. Herzog, A. Simis, W.V. Vasconcelos, On the arithmetic and homology of algebras of linear type. Trans. Am. Math. Soc. **283**, 661–683 (1984)
18. J. Hong, A. Simis, W.V. Vasconcelos, On the homology of two-dimensional elimination. J. Symbolic Comput. **43**, 275–292 (2008)
19. C. Huneke, On the symmetric algebra and Rees algebra of an ideal generated by a d-sequence. J. Algebra **62**, 268–275 (1980)
20. C. Huneke, On the associated graded algebra of an ideal. Ill. J. Math. **26**, 121–137 (1982)
21. C. Huneke, Linkage and the Koszul homology of ideals. Am. J. Math. **104**, 1043–1062 (1982)

22. C. Huneke, Determinantal ideals of linear type. Arch. Math. **47**, 324–329 (1986)
23. C. Huneke, M. Rossi, The dimension and components of symmetric algebras. J. Algebra **98**, 200–210 (1986)
24. J.-P. Jouanolou, Formes d'inertie et resultant: un formulaire. Adv. Math. **126**, 119–250 (1997)
25. V. Kodiyalam, Castelnuovo-Mumford regularity. Proc. Am. Math. Soc. **128**, 407–411 (2000)
26. B.V. Kotzev, Determinantal ideals of linear type of a generic symmetric matrix. J. Algebra **139**, 484–504 (1991)
27. A. Kustin, B. Ulrich, A family of complexes associated to an almost alternating map, with applications to residual intersections. Mem. Am. Math. Soc. **95**(461), 1–95 (1992)
28. A. Kustin, C. Polini, B. Ulrich, Divisors on rational normal scrolls. J. Algebra **322**, 1748–1773 (2009)
29. A. Kustin, C. Polini, B. Ulrich, Rational normal scrolls and the defining equations of Rees algebras. J. Reine Angew. Math. **650**, 23–65 (2011)
30. A. Kustin, C. Polini, B. Ulrich, Blowups and fibers of morphisms. Nagoya Math. J. **224**, 168–201 (2016)
31. A. Kustin, C. Polini, B. Ulrich, The Hilbert series of the ring associated to an almost alternating matrix. Comm. Algebra **44**, 3053–3068 (2016)
32. A. Kustin, C. Polini, B. Ulrich, The bi-graded structure of symmetric algebras with applications to Rees rings. J. Algebra **469**, 188–250 (2017)
33. A. Kustin, C. Polini, B. Ulrich, Degree bounds for local cohomology. Preprint (2017)
34. A. Kustin, C. Polini, B. Ulrich, The equations defining blowup algebras of height three Gorenstein ideals. Algebra Number Theory **11**, 1489–1525 (2017)
35. A. Kustin, C. Polini, B. Ulrich, A matrix of linear forms which is annihilated by a vector of indeterminates. J. Algebra **469**, 120–187 (2017)
36. N. Lan, On Rees algebras of linearly presented ideals. J. Algebra **420**, 186–200 (2014)
37. N. Lan, On Rees algebras of linearly presented ideals in three variables. J. Pure Appl. Algebra **221**, 2180–2191 (2017)
38. K.-N. Lin, Rees algebras of diagonal ideals. J. Comm. Algebra **5**, 359–398 (2013)
39. K.-N. Lin, C. Polini, Rees algebras of truncations of complete intersections. J. Algebra **410**, 36–52 (2014)
40. J. Madsen, Equations of Rees algebras and singularities of rational plane curves. Ph.D. Thesis, University of Notre Dame, 2016
41. J. Madsen, Equations of Rees algebras of ideals in two variables. Preprint (2016)
42. H. Matsmura, *Commutative Ring Theory*. Cambridge Studies in Advanced Mathematics, vol. 8 (Cambridge University Press, Cambridge, 1989), xiv-320
43. A. Micali, Algèbres intègres et sans torsion. Bull. Soc. Math. France **94**, 5–13 (1966)
44. S. Morey, B. Ulrich, Rees algebras of ideals with low codimension. Proc. Am. Math. Soc. **124**, 3653–3661 (1996)
45. C. Peskine, L. Szpiro, Dimension projective finie et cohomologie locale. Applications à la démonstration de conjectures de M. Auslander, H. Bass et A. Grothendieck. Inst. Hautes Études Sci. Publ. Math. **42**, 47–119 (1973)
46. F. Schreyer, Syzygies of canonical curves and special linear series. Math. Ann. **275**, 105–137 (1986)
47. A. Simis, W.V. Vasconcelos, The syzygies of the conormal module. Am. J. Math. **103**, 203–224 (1981)
48. A. Simis, W.V. Vasconcelos, On the dimension and integrality of symmetric algebras. Math. Z. **177**, 341–358 (1981)
49. N.V. Trung, H.-J. Wang, On the asymptotic linearity of Castelnuovo-Mumford regularity. J. Pure Appl. Algebra **201**, 42–48 (2005)
50. B. Ulrich, Ideals having the expected reduction number. Am. J. Math. **118**, 17–38 (1996)
51. G. Valla, On the symmetric and Rees algebras of an ideal. Manuscripta Math. **30**, 239–255 (1980)
52. W.V. Vasconcelos, On the equations of Rees algebras. J. Reine Angew. Math. **418**, 189–218 (1991)

53. J. Watanabe, A note on Gorenstein rings of embedding codimension three. Nagoya Math. J. **50**, 227–332 (1973)
54. J. Weyman, Resolutions of the exterior and symmetric powers of a module. J. Algebra **58**, 333–341 (1979)

Chapter 2
Koszul Modules

Claudiu Raicu

Abstract The goal of these notes is to introduce the basic theory of Koszul modules, and survey some applications to syzygies and to Green's conjecture, following recent work of Aprodu, Farkas, Papadima, Raicu, and Weyman (Invent. Math. **218**(3), 657–720 (2019)). More precisely, we discuss a relationship between Koszul modules and the syzygies of the tangent developable surface to a rational normal curve, which arises from a version of Hermite reciprocity for SL_2-representations. We describe in terms of the characteristic of the underlying field which of the Betti numbers of this surface are zero and which ones are not. We indicate how, by passing to a hyperplane section and using degenerations, the generic version of Green's conjecture can be deduced in almost all characteristics. These notes are accompanied by examples and exercises, and include an Appendix on multilinear algebra.

2.1 Syzygies of the Tangential Variety to a Rational Normal Curve

Throughout this chapter we work over an algebraically closed field $\mathbf{k}$. We write $\mathbf{P}^r$ for the r-dimensional projective space over $\mathbf{k}$. Consider the **rational normal curve of degree** g, denoted Γ_g and defined as the image of the **Veronese map**

$$v_g : \mathbf{P}^1 \longrightarrow \mathbf{P}^g, \quad [a : b] \longrightarrow [a^g : a^{g-1}b : \cdots : ab^{g-1} : b^g]. \tag{2.1}$$

As with all parametrized varieties, a first natural question to consider is the **implicitization problem** of finding the ideal of equations that vanish along Γ_g

C. Raicu (✉)
University of Notre Dame , Notre Dame, IN, USA

Institute of Mathematics "Simion Stoilow" of the Romanian Academy , Bucharest, Romania
e-mail: craicu@nd.edu

© The Editor(s) (if applicable) and The Author(s), under exclusive license
to Springer Nature Switzerland AG 2021
A. Conca et al. (eds.), *Recent Developments in Commutative Algebra*, Lecture
Notes in Mathematics 2283, https://doi.org/10.1007/978-3-030-65064-3_2

[3, Section 3.3]. To answer it, we let $S = \mathbf{k}[z_0, \cdots, z_g]$ denote the homogeneous coordinate ring of $\mathbf{P}^g$, let $A = \mathbf{k}[x, y]$ denote the homogeneous coordinate ring of $\mathbf{P}^1$, and consider the pull-back homomorphism

$$\phi: S \longrightarrow A, \quad \phi(z_i) = x^{g-i}y^i, \text{ for } i = 0, \cdots, g.$$

The ideal $I(\Gamma_g)$ of polynomials vanishing along Γ_g is equal to $\ker(\phi)$. Since the matrix

$$\begin{bmatrix} x^g & x^{g-1}y & \cdots & xy^{g-1} \\ x^{g-1}y & x^{g-2}y^2 & \cdots & y^g \end{bmatrix}$$

has proportional rows (with ratio x/y), it follows that its 2×2 minors vanish identically. In other words, the 2×2 minors of

$$Z = \begin{bmatrix} z_0 & z_1 & \cdots & z_{g-1} \\ z_1 & z_2 & \cdots & z_g \end{bmatrix} \tag{2.2}$$

belong to $\ker(\phi) = I(\Gamma_g)$. For the following, you may consult [6, Proposition 6.1].

Exercise 2.1 Show that $I(\Gamma_g)$ is generated by the 2×2 minors of Z.

The next natural step is to investigate the minimal free resolution of the homogeneous coordinate ring $S/I(\Gamma_g)$ of Γ_g (see [6] for a comprehensive introduction to the study of minimal resolutions in Algebraic Geometry). We let

$$B_{i,j}(\Gamma_g) = \mathrm{Tor}_i^S(S/I(\Gamma_g), \mathbf{k})_{i+j}$$

denote the module of i-**syzygies of weight** j (or degree $i + j$) of Γ_g, and define the **Betti numbers of** Γ_g via

$$b_{i,j}(\Gamma_g) = \dim_{\mathbf{k}} B_{i,j}(\Gamma_g).$$

The Betti numbers of Γ_g are recorded into the **Betti table** $\beta(\Gamma_g)$, where the columns account for the homological degree, and the rows for internal degree:

$$
\begin{array}{c|ccc}
 & & i & \\
\hline
 & & \vdots & \\
j & \cdots & b_{i,j}(\Gamma_g) & \cdots \\
 & & \vdots &
\end{array}
$$

For example, when $g = 3$, Γ_g is known as the **twisted cubic curve**, with Betti table

$$
\begin{array}{c|ccc}
 & 0 & 1 & 2 \\
\hline
0 & 1 & - & - \\
1 & - & 3 & 2
\end{array}
$$

where a dash indicates the vanishing of the corresponding Betti number. We say that the twisted cubic has a **linear minimal free resolution** (after the first step), since all (but the first) Betti numbers are concentrated in a single row. This is true more generally for any Γ_g, whose minimal free resolution is given by an Eagon–Northcott complex (see Appendix), and the corresponding Betti table takes the following shape:

$$
\begin{array}{c|cccccccc}
 & 0 & 1 & 2 & \cdots & i & \cdots & g-1 \\
\hline
0 & 1 & - & - & \cdots & - & \cdots & - \\
1 & - & \binom{g}{2} & 2 \cdot \binom{g}{3} & \cdots & i \cdot \binom{g}{i+1} & \cdots & (g-1)
\end{array}
$$

Remark 2.1 The description of the defining equations and that of the Betti table of Γ_g is independent on the characteristic of the field **k**.

Let $\mathcal{T}_g$ denote the **tangential variety (or tangent developable) of Γ_g**, defined as the union of the tangent lines to Γ_g. One of the problems that we will be concerned with in these notes is the following.

Problem 2.1 Describe the defining equations and the Betti table of $\mathcal{T}_g$.

The main result in this direction will be a description of the (non-)vanishing behavior of $b_{i,j}(\mathcal{T}_g)$ (see Theorem 2.1). Based on this description, finding the actual value of the Betti numbers is fairly easy in characteristic 0 or sufficiently large, but remains unknown in small characteristics.

2.1.1   The Local Description of Γ_g, $\mathcal{T}_g$

Restricting the Veronese map v_g in (2.1) to the affine chart $x = 1$ yields a local parametrization of Γ_g via

$$
t \mapsto [1 : t : \cdots : t^g] = (t, t^2, \cdots, t^g),
$$

where [$\cdots$] represents projective notation, and ($\cdots$) represents affine notation. The tangent directions to Γ_g are determined by differentiating with respect to t, so $\mathcal{T}_g$ is described in the affine chart $z_0 = 1$ by the parametrization

$$(t, s) \longrightarrow (t, t^2, \cdots, t^g) + s \cdot (1, 2t, \cdots, gt^{g-1}) = (t+s, t^2+2ts, \cdots, t^g+gt^{g-1}s). \tag{2.3}$$

Since $\mathcal{T}_g$ is irreducible, one can use the local description above to check when a homogeneous polynomial belongs to $I(\mathcal{T}_g)$. This observation can be applied to check that the quadratic equations constructed below vanish along $\mathcal{T}_g$. Let

$$\Delta_{i,j} = \det \begin{bmatrix} z_i & z_j \\ z_{i+1} & z_{j+1} \end{bmatrix}, \text{ for } 0 \leq i, j \leq g - 1,$$

and define

$$Q_{i,j} = \Delta_{i+2,j} - 2 \cdot \Delta_{i+1,j+1} + \Delta_{i,j+2}, \text{ for } 0 \leq i < j \leq g - 3. \tag{2.4}$$

Exercise 2.2 Check that $Q_{i,j} \in I(\mathcal{T}_g)$ for all $0 \leq i < j \leq g - 3$.

Remark 2.2 As we'll see later, the quadrics constructed above will generate, in most cases, the ideal $I(\Gamma_g)$. The exceptions are when $g \leq 4$, or when char($\mathbf{k}$) $\in \{2, 3\}$.

2.1.2 Characteristic Zero Interpretation of Γ_g, $\mathcal{T}_g$

Suppose that char($\mathbf{k}$) $= 0$ (or more generally, that the binomial coefficients $\binom{g}{i}$, $i = 1, \cdots, g-1$, are invertible in $\mathbf{k}$), and consider the vector space of homogeneous forms of degree g in the variables X, Y. Since the binomial coefficients are invertible in $\mathbf{k}$, one can write each such form uniquely as

$$F(X, Y) = z_0 \cdot X^g + z_1 \cdot \binom{g}{1} \cdot X^{g-1}Y + \cdots + z_i \cdot \binom{g}{i} \cdot X^{g-i}Y^i + \cdots + z_g \cdot Y^g,$$

for $z_0, \cdots, z_g \in \mathbf{k}$. Notice that $F(X, Y)$ is a power of a linear form, namely $F(X, Y) = (aX + bY)^g$, if and only if

$$z_0 = a^g, \ z_1 = a^{g-1}b, \cdots, \ z_i = a^{g-i}b^i, \cdots, \ z_g = b^g.$$

It follows that Γ_g parametrizes (up to scaling) binary forms $F(X, Y) \neq 0$ that factor as a power L^g of a linear form L in X, Y.

Exercise 2.3 Check that $\mathcal{T}_g$ parametrizes (up to scaling) binary forms $F(X, Y) \neq 0$ that factor as a product $L_1^{g-1} \cdot L_2$, where L_1, L_2 are linear forms in X, Y.

In the case $g = 3$, it follows that $\mathcal{T}_3$ parametrizes cubic forms with a double root (in $\mathbf{P}^1$). This is a quartic surface in $\mathbf{P}^3$, defined by the vanishing of the discriminant of F (see [12, Example 2.22]):

$$- 3z_1^2 z_2^2 + 4z_0 z_2^3 + 4z_1^3 z_3 - 6z_0 z_1 z_2 z_3 + z_0^2 z_3^2 = 0. \tag{2.5}$$

If char($\mathbf{k}$) = 2, this equation becomes $(z_0 z_3 - z_1 z_2)^2 = 0$, $I(\mathcal{T}_3) = \langle z_0 z_3 - z_1 z_2 \rangle$, and $\mathcal{T}_3 \simeq \mathbf{P}^1 \times \mathbf{P}^1$ is a quadric surface in $\mathbf{P}^3$. The following section describes more generally the situation in characteristic 2.

2.1.3 The Betti Table of $\mathcal{T}_g$ in Characteristic 2

Suppose char($\mathbf{k}$) = 2 and consider the matrix of linear forms

$$M = \begin{bmatrix} z_0 & z_1 & \cdots & z_{g-2} \\ z_2 & z_3 & \cdots & z_g \end{bmatrix}. \tag{2.6}$$

We claim that the 2×2 minors of M vanish on $\mathcal{T}_g$. As noted earlier, it suffices to check this assertion on the chart (2.3). Evaluating M on the point of $\mathcal{T}_g$ parametrized by (t, s) we obtain

$$M(t, s) = \begin{bmatrix} 1 & t & t^2 & t^3 & \cdots & t^{g-2} \\ t^2 & t^3 & t^4 & t^5 & \cdots & t^g \end{bmatrix} + s \cdot \begin{bmatrix} 0 & t & 0 & t^3 & \cdots & (g-2) \cdot t^{g-2} \\ 0 & t^3 & 0 & t^5 & \cdots & g \cdot t^g \end{bmatrix}$$

Since $(i - 2) = i$ for all $i \in \mathbf{k}$, the second row of $M(t, s)$ is obtained from the first by multiplying by t^2, so $M(t, s)$ has rank (at most) one. The 2×2 minors of M define a rational normal scroll S of dimension 2 (see the Appendix, or [6, p.107]), so that $\mathcal{T}_g \subseteq S$. Since $\dim(\mathcal{T}_g) = 2$, it follows that $\mathcal{T}_g = S$. It follows that the minimal resolution of $\mathcal{T}_g$ is given by an Eagon–Northcott complex (see the Appendix, or [6, Corollary A2.62]), and the Betti table takes the form

	0	1	2	$\cdots$	i	$\cdots$	$g-2$
0	1	$-$	$-$	$\cdots$	$-$	$\cdots$	$-$
1	$-$	$\binom{g-1}{2}$	$2 \cdot \binom{g-1}{3}$	$\cdots$	$i \cdot \binom{g-1}{i+1}$	$\cdots$	$(g-2)$

Note that $I(\mathcal{T}_g)$ is generated in this case by $\binom{g-1}{2}$ quadrics, so that the $\binom{g-2}{2}$ quadrics constructed in (2.4) are not enough to generate $I(\mathcal{T}_g)$.

2.1.4   The Betti Numbers of $\mathcal{T}_g$ for char(k) $\neq 2$

If char(k) $\neq 2$, it is shown in [2, Theorem 5.1] that the homogeneous coordinate ring of $\mathcal{T}_g$ is Gorenstein of Castelnuovo–Mumford regularity 3, that is, the Betti table of $\mathcal{T}_g$ has the following shape:

$$
\begin{array}{c|ccccccc}
 & 0 & 1 & 2 & \cdots & g-4 & g-3 & g-2 \\
\hline
0 & 1 & - & - & \cdots & - & - & - \\
1 & - & b_{1,1} & b_{2,1} & \cdots & b_{g-4,1} & b_{g-3,1} & - \\
2 & - & b_{1,2} & b_{2,2} & \cdots & b_{g-4,2} & b_{g-3,2} & - \\
3 & - & - & - & \cdots & - & - & 1
\end{array}
\tag{2.7}
$$

with $b_{i,1} = b_{g-2-i,2}$ for $i = 1, \cdots, g-3$. In order to understand the Betti table, it is then sufficient to analyze the row $b_{\bullet,2}$. The following theorem, whose proof we'll outline later, completely characterizes the (non-)vanishing behavior of the Betti numbers of $\mathcal{T}_g$ [2, Theorem 5.2].

Theorem 2.1 *Suppose that $p = $ char(k) $\neq 2$. If $p = 0$ or $p \geq \frac{g+2}{2}$, then*

$$
b_{i,2}(\mathcal{T}_g) \neq 0 \quad \Longleftrightarrow \quad \frac{g-2}{2} \leq i \leq g - 3.
\tag{2.8}
$$

If $3 \leq p \leq \frac{g+1}{2}$, then

$$
b_{i,2}(\mathcal{T}_g) \neq 0 \quad \Longleftrightarrow \quad p - 2 \leq i \leq g - 3.
\tag{2.9}
$$

Notice that taking $p = 3$ and $g \geq 5$ in Theorem 2.1 yields $b_{1,2}(\mathcal{T}_g) \neq 0$, so the ideal $I(\mathcal{T}_g)$ requires cubic minimal generators. If $g = 3$ then $\mathcal{T}_g$ is a quartic surface, while for $g = 4$ it is the intersection of a quadric and a cubic threefold. In all other cases, $\mathcal{T}_g$ is cut out by quadratic equations.

Exercise 2.4 Show that if $p \neq 2$ then $b_{1,1}(\mathcal{T}_g) = \binom{g-2}{2}$, and (2.4) gives all the quadrics vanishing on $\mathcal{T}_g$.

Example 2.1 To illustrate the dependence of the Betti numbers of $\mathcal{T}_g$ on the characteristic, consider the case when $g = 5$. If $p \neq 2, 3$, it follows from Exercise 2.4 and Theorem 2.1 that $\beta(\mathcal{T}_5)$ is

$$
\begin{array}{c|cccc}
 & 0 & 1 & 2 & 3 \\
\hline
0 & 1 & - & - & - \\
1 & - & 3 & - & - \\
2 & - & - & 3 & - \\
3 & - & - & - & 1
\end{array}
$$

so that $\mathcal{T}_5$ is a complete intersection of three quadrics. If $p = 3$ then it can be checked that the Betti table is

	0	1	2	3
0	1	–	–	–
1	–	3	2	–
2	–	2	3	–
3	–	–	–	1

while for $p = 2$, we have seen that $\beta(\mathcal{T}_5)$ is

	0	1	2	3
0	1	–	–	–
1	–	6	8	3

There is one part of Theorem 2.1 that can be checked using ideas from Sect. 2.1.3:

Exercise 2.5 Show that if $p < g$ then the 2×2 minors of

$$\begin{bmatrix} z_0 & z_1 & \cdots & z_{g-p} \\ z_p & z_{p+1} & \cdots & z_g \end{bmatrix} \tag{2.10}$$

vanish on $\mathcal{T}_g$. Show that these minors cut out a scroll S of dimension p, and conclude that $b_{i,1}(S) \neq 0$ for $1 \leq i \leq g - p$. Deduce the corresponding non-vanishing for $b_{i,1}(\mathcal{T}_g)$, and derive (using the Gorenstein property) the implication "$\Longleftarrow$" in (2.9).

The next exercises indicate how to compute the Betti numbers of $\mathcal{T}_g$ using Theorem 2.1, provided that $p = 0$ or $p \geq \frac{g+2}{2}$.

Exercise 2.6 (See [15, Section 6]) If $p \neq 2$, show that the Hilbert series of $\mathcal{T}_g$ is given by

$$\sum_{d \geq 0} \dim(S/I(\mathcal{T}_g))_d \cdot t^d = \frac{1 + (g - 2)t + (g - 2)t^2 + t^3}{(1 - t)^3}.$$

Conclude that $\mathcal{T}_g$ is a surface of degree $2g - 2$.

Using the fact that the alternating sums along antidiagonals in a Betti table are determined by the Hilbert function [6, Corollary 1.10], show the following.

Exercise 2.7 Suppose that $p = 0$ or $p \geq \frac{g+2}{2}$. The non-zero Betti numbers $b_{\bullet,1}(\mathcal{T}_g)$ are computed as follows:

$$b_{i,1}(\mathcal{T}_g) = i \cdot \binom{g-2}{i+1} - (g - 1 - i) \cdot \binom{g-2}{i-2} \text{ for } i = 1, \cdots, \left\lfloor \frac{g-2}{2} \right\rfloor.$$

In characteristic $p = 2$ we have seen how to compute the Betti numbers of $\mathcal{T}_g$ in Sect. 2.1.3. In other small characteristics, these numbers are not known.

Problem 2.2 (Open) When $3 \leq p \leq \frac{g+1}{2}$, find a formula for the Betti numbers of $\mathcal{T}_g$.

2.2 Koszul Modules

Suppose that V is an n-dimensional **k**-vector space and fix a subspace $K \subseteq \bigwedge^2 V$ with $\dim(K) = m$. We denote by $S := \mathrm{Sym}(V)$ the symmetric algebra over V and consider the Koszul complex resolving the residue field **k**:

$$\cdots \longrightarrow \overset{3}{\bigwedge} V \otimes S \xrightarrow{\delta_3} \overset{2}{\bigwedge} V \otimes S \xrightarrow{\delta_2} V \otimes S \xrightarrow{\delta_1} S.$$

Truncating this complex to the last three terms, and restricting δ_2 along the inclusion $\iota \colon K \hookrightarrow \bigwedge^2 V$ we obtain a 3-term complex

$$K \otimes S \xrightarrow{\delta_2|_{K \otimes S}} V \otimes S(1) \xrightarrow{\delta_1} S(2). \tag{2.11}$$

The **Koszul module** associated to the pair (V, K) is the middle homology of the complex (2.11). We make the convention that K is placed in degree zero, so that $W(V, K)$ is a graded S-module generated in degree zero. In particular, the degree q component of $W(V, K)$ is given by

$$W_q(V, K) = \text{middle homology of } \left(K \otimes \mathrm{Sym}^q V \longrightarrow V \otimes \mathrm{Sym}^{q+1} V \longrightarrow \mathrm{Sym}^{q+2} V \right)$$

The formation of the Koszul module $W(V, K)$ is natural in the following sense. An inclusion $K \subseteq K'$ induces a surjective morphism of graded S-modules

$$W(V, K) \twoheadrightarrow W(V, K'), \tag{2.12}$$

that is, *bigger* subspaces $K \subseteq \bigwedge^2 V$ correspond to *smaller* Koszul modules. For instance, we have that $W(V, K) = 0$ if and only if $K = \bigwedge^2 V$. We'll be interested more generally in studying Koszul modules of finite length, that is, those that satisfy $W_q(V, K) = 0$ for $q \gg 0$. Since $W(V, K)$ is generated in degree zero, the vanishing $W_q(V, K) = 0$ for some $q \geq 0$ implies that $W_{q'}(V, K) = 0$ for all $q' \geq q$.

We write $\iota^\vee \colon \bigwedge^2 V^\vee \twoheadrightarrow K^\vee$ for the dual to the inclusion ι, let $K^\perp := \ker(\iota^\vee) \subseteq \bigwedge^2 V^\vee$ and define the *resonance variety* $\mathcal{R}(V, K)$ by

$$\mathcal{R}(V, K) := \left\{ a \in V^\vee : \text{ there exists } b \in V^\vee \text{ such that } a \wedge b \in K^\perp \setminus \{0\} \right\} \cup \{0\}. \tag{2.13}$$

The following result is proved in [13, Lemma 2.4].

Lemma 2.1 *The resonance variety $\mathcal{R}(V, K)$ coincides with the set-theoretic support of $W(V, K)$ in the affine space $V^\vee$.*

Proof We let $\mathbf{P} = \mathbb{P}V^\vee$ denote the projective space of one dimensional subspaces of $V^\vee$, and consider the **Koszul sheaf** $\mathcal{W}(V, K)$, defined by considering the complex of sheaves associated to (2.11): $\mathcal{W}(V, K)$ is the middle homology of

$$K \otimes O_\mathbf{P} \longrightarrow V \otimes O_\mathbf{P}(1) \longrightarrow O_\mathbf{P}(2).$$

Since $W(V, K)$ is a graded module, its set-theoretic support is the affine cone over the support of $\mathcal{W}(V, K)$.

If we write $\Omega = \Omega_\mathbf{P}^1$ for the sheaf of differential forms, then the Euler sequence

$$0 \longrightarrow \Omega \longrightarrow V \otimes O_\mathbf{P}(-1) \longrightarrow O_\mathbf{P} \longrightarrow 0 \tag{2.14}$$

yields the identification $\ker(V \otimes O_\mathbf{P}(1) \longrightarrow O_\mathbf{P}(2)) = \Omega(2)$, so that

$$\mathcal{W}(V, K) = \operatorname{coker}(K \otimes O_\mathbf{P} \longrightarrow \Omega(2)).$$

It follows that the support of $\mathcal{W}(V, K)$ is the locus where the map $K \otimes O_\mathbf{P} \longrightarrow \Omega(2)$ fails to be surjective. By Nakayama's lemma, this is a condition that can be checked on fibers.

Consider a point $p = [f] \in \mathbf{P}$, where $0 \neq f \in V^\vee$. The restriction of (2.14) to the fiber at p is identified with

$$0 \longrightarrow \ker(f) \longrightarrow V \xrightarrow{f} \mathbf{k} \longrightarrow 0 , \tag{2.15}$$

so the restriction of the map $K \otimes O_\mathbf{P} \longrightarrow \Omega(2)$ to the fiber at p is given by the contraction by f map $K \longrightarrow \ker(f)$. It follows that p is in the support of $\mathcal{W}(V, K)$ if and only if the corresponding sequence

$$K \longrightarrow V \xrightarrow{f} \mathbf{k}$$

fails to be exact. This is equivalent to the failure of exactness of the dual sequence

$$\mathbf{k} \xrightarrow{\cdot f} V^\vee \xrightarrow{\wedge f} K^\vee, \tag{2.16}$$

where the second map is the composition $V^\vee \xrightarrow{\wedge f} \bigwedge^2 V^\vee \twoheadrightarrow K^\vee$. It follows that a cycle in (2.16) is an element $g \in V^\vee$ with $g \wedge f \in K^\perp$, and g gives a non-trivial homology class if and only if g is not a multiple of f, that is, if $g \wedge f \neq 0$. Using (2.13) the existence of such g is equivalent to the fact that $f \in \mathcal{R}(V, K)$, which concludes our proof.   □

It follows from Lemma 2.1 that $W(V, K)$ has finite length if and only $\mathcal{R}(V, K) = \{0\}$. In view of (2.13), this last condition is equivalent to the fact that the linear subspace $\mathbb{P}K^\perp \subseteq \mathbb{P}(\bigwedge^2 V^\vee)$ is disjoint from the Grassmann variety (of 2-planes in $V^\vee$)

$$\mathbf{G} := \mathrm{Gr}_2(V^\vee)$$

in its Plücker embedding. This can happen only when

$$m = \mathrm{codim}(\mathbb{P}K^\perp) > \dim(\mathbf{G}) = 2n - 4.$$

Summarizing, we have the following equivalences:

$$\mathbb{P}K^\perp \cap \mathbf{G} = \emptyset \iff \mathcal{R}(V, K) = \{0\} \iff \dim_{\mathbf{k}} W(V, K) < \infty. \tag{2.17}$$

Moreover, if the equivalent statements in (2.17) hold, then $m \geq 2n - 3$. The following theorem gives a sharp vanishing result for the graded components of a Koszul module with vanishing resonance [2, Theorem 1.3].

Theorem 2.2 *Suppose that $n \geq 3$. If $\mathrm{char}(\mathbf{k}) = 0$ or $\mathrm{char}(\mathbf{k}) \geq n - 2$, then we have the equivalence*

$$\mathcal{R}(V, K) = \{0\} \iff W_q(V, K) = 0 \text{ for } q \geq n - 3. \tag{2.18}$$

Exercise 2.8 Check Theorem 2.2 in the case when $n = 3$.

Exercise 2.9 Show that if $\mathcal{R}(V, K) = \{0\}$ then there exists a subspace $K' \subseteq K$ with $\dim(K') = 2n - 3$ such that $\mathcal{R}(V, K') = \{0\}$. Conclude that the implication "$\Longrightarrow$" in (2.18) reduces to the case when $\dim(K) = 2n - 3$.

Proof of Theorem 2.2 The implication "$\Longleftarrow$" follows from (2.17). To prove "$\Longrightarrow$", we assume that (V, K) is such that $\mathcal{R}(V, K) = \{0\}$ and $\dim(K) = 2n - 3$. With notation as in the proof of Lemma 2.1, it follows that the natural map $\alpha: K \otimes O_{\mathbb{P}} \longrightarrow \Omega(2)$ is surjective, and therefore it gives rise to an exact Buchsbaum–Rim complex $\mathcal{B}_\bullet = \mathbf{BR}_\bullet(\alpha)$ with

$$\mathcal{B}_0 = \Omega(2), \quad \mathcal{B}_1 = K \otimes O_{\mathbb{P}},$$

$$\mathcal{B}_r = \bigwedge^{n+r-2} K \otimes \det\left(\Omega^\vee(-2)\right) \otimes \mathrm{D}^{r-2}\left(\Omega^\vee(-2)\right)$$

$$= \bigwedge^{n+r-2} K \otimes O(-n-2r+6) \otimes \mathrm{D}^{r-2}(\Omega^\vee) \text{ for } r = 2, \cdots, n-1.$$

The condition $W_{n-3}(V, K) = 0$ is equivalent to the fact that after twisting by $O_{\mathbf{P}}(n-3)$, the induced map on global sections

$$H^0(\mathbf{P}, \mathcal{B}_1(n-3)) \longrightarrow H^0(\mathbf{P}, \mathcal{B}_0(n-3)) \tag{2.19}$$

is surjective. Since $\mathcal{B}_\bullet(n-3)$ is an exact complex, its hypercohomology groups are all zero. Using the hypercohomology spectral sequence, in order to prove the surjectivity of (2.19) it suffices to check that the sheaves $\mathcal{B}_r(n-3)$ have no cohomology (in fact, it is enough that $H^{r-1}(\mathbf{P}, \mathcal{B}_r(n-3)) = 0$) for $r = 2, \cdots, n-1$. Since $0 \leq r-2 \leq n-3$, it follows from our hypothesis that $p = \mathrm{char}(\mathbf{k})$ satisfies $p = 0$ or $p > r-2$, thus $\mathrm{D}^{r-2}(\Omega^\vee) = \mathrm{Sym}^{r-2}(\Omega^\vee)$. It follows that

$$\mathcal{B}_r(n-3) = \bigwedge^{n+r-2} K \otimes \mathrm{Sym}^{r-2}\left(\Omega^\vee\right) \otimes O(-2r+3), \text{ for } r = 2, \cdots, n-1,$$

and it suffices to check that $\mathrm{Sym}^{r-2}(\Omega^\vee) \otimes O(-2r+3)$ has no non-zero cohomology for $r = 2, \cdots, n-1$. Dualizing the Euler sequence (2.14), taking symmetric powers, and twisting by $O(-2r+3)$, we obtain a short exact sequence

$$0 \longrightarrow \mathrm{Sym}^{r-3}(V^\vee) \otimes O(-r) \longrightarrow \mathrm{Sym}^{r-2}(V^\vee) \otimes O(-r+1)$$
$$\longrightarrow \mathrm{Sym}^{r-2}(\Omega^\vee) \otimes O(-2r+3) \longrightarrow 0.$$

It is then enough to check that $O(-r)$ and $O(-r+1)$ have no non-zero cohomology when $r = 2, \cdots, n-1$, which follows from the fact that $-n < -r, -r+1 < 0$ and cohomology of line bundles on projective space vanishes in this range. $\square$

Remark 2.3 If you know about Castelnuovo-Mumford regularity, then you can replace the spectral sequence argument in the proof above with the following (which I learned from Rob Lazarsfeld). Dualizing the Euler sequence (2.14) we get that $\Omega^\vee(-1)$ has a two-step resolution $0 \longrightarrow O(-1) \longrightarrow V^\vee \otimes O$. Since O is 0-regular and $O(-1)$ is 1-regular, we conclude that $\Omega^\vee(-1)$ is 0-regular. A tensor product of copies of $\Omega^\vee(-1)$ will then also be 0-regular. Under our assumptions, $\mathrm{D}^{r-2}(\Omega^\vee) \otimes O(-r+2) = \mathrm{D}^{r-2}(\Omega^\vee(-1))$ is a direct summand in a tensor product of $(r-2)$ copies of $\Omega^\vee(-1)$, so it is itself 0-regular. Since $O(-i)$ is i-regular for all i, it follows that $\mathcal{B}_r(n-3)$ is $(r-1)$-regular for $r \geq 2$. If we let $\mathcal{J} = \ker(\mathcal{B}_1(n-3) \longrightarrow \mathcal{B}_0(n-3))$ it follows that $\mathcal{J}$ has a resolution $\mathcal{B}_{\bullet \geq 2}(n-3)$, where the i-th term $\mathcal{B}_{i+2}(n-3)$ is $(i+1)$-regular. This implies that $\mathcal{J}$ is 1-regular, so that $H^1(\mathbf{P}, \mathcal{J}) = 0$. From the long exact sequence

$$\cdots \longrightarrow H^0(\mathbf{P}, \mathcal{B}_1(n-3)) \longrightarrow H^0(\mathbf{P}, \mathcal{B}_0(n-3)) \longrightarrow H^1(\mathbf{P}, \mathcal{J}) \longrightarrow \cdots$$

it follows that the map (2.19) is surjective, as desired.

Experimental evidence suggests that $\mathrm{char}(\mathbf{k}) \geq n-2$ is the precise hypothesis necessary for (2.18) to hold. Similarly, the vanishing range $q \geq n-3$ is optimal, as shown by the following [2, Theorem 1.4].

Theorem 2.3 *Suppose* $\mathrm{char}(\mathbf{k}) = 0$ *or* $\mathrm{char}(\mathbf{k}) \geq n-2$, *and fix a subspace* $K \subseteq \bigwedge^2 V$. *If* $\mathcal{R}(V, K) = \{0\}$, *then*

$$\dim W_q(V, K) \leq \binom{n+q-1}{q} \frac{(n-2)(n-q-3)}{q+2} \quad \text{for } q = 0, \ldots, n-4.$$

Moreover, equality holds for all q if $\dim(K) = 2n-3$.

The table below records some of the values of $\dim W_q(V, K)$ in the case when equality holds in Theorem 2.3.

$q \backslash^n$	4	5	6	7	8	
0	1	3	6	10	15	
1	–	5	16	35	64	
2	–	–	21	70	162	(2.20)
3	–	–	–	84	288	
4	–	–	–	–	330	

Exercise 2.10 Find a formula for $\dim(K \otimes \mathrm{Sym}^q V)$ and $\dim(W_q(V, 0))$, and check that if $\dim(K) = 2n-3$ then

$$\dim(W_q(V, 0)) - \dim(K \otimes \mathrm{Sym}^q V) = \binom{n+q-1}{q} \frac{(n-2)(n-q-3)}{q+2}.$$

Proof of Theorem 2.3 Using (2.12) and Exercise 2.9, we can reduce to the case when $\dim(K) = 2n-3$. We have that $W_q(V, K)$ is the cokernel of the natural map

$$\beta_q : K \otimes \mathrm{Sym}^q V \longrightarrow W_q(V, 0).$$

When $q = n-3$, the source and target have the same dimension. By Theorem 2.2, $W_{n-3}(V, K) = 0$, so β_{n-3} is an isomorphism, and in particular it is injective. Since $\beta = \bigoplus_q \beta_q : K \otimes S \longrightarrow W(V, 0)$ is a map of S-modules, whose source is free, it follows that the injectivity of β_{n-3} implies that of β_q for all $q \leq n - 3$. This shows that

$$\dim(W_q(V, K)) = \dim(W_0(V, K)) - \dim(K \otimes \operatorname{Sym}^q V) \text{ for } q = 0, \cdots, n - 3,$$

and the desired formula follows from Exercise 2.10.   □

Example 2.2 Examples computed with Macaulay2 [9] show that Theorem 2.2 may fail in small characteristics. For instance, when $n = 6$ and char$(\mathbf{k}) = 3$, there are examples of Koszul modules with Hilbert function

q	0	1	2	3	4	5	$\cdots$
$\dim W_q(V, K)$	6	16	21	6	1	0	$\cdots$

We expect that finding the appropriate extensions of the previous results in small characteristics is closely related to understanding the failure of the Borel–Weil–Bott theorem in the said characteristics (see also [2, Section 2]).

2.3 Weyman Modules, Hermite Reciprocity, and the Syzygies of $\mathcal{T}_g$

In this section we introduce a natural class of examples of Koszul modules that satisfy (in most characteristics) the hypotheses of Theorems 2.2 and 2.3. They are called **Weyman modules**, following [4, Section 3.I.B], and via Hermite reciprocity will provide a link with the Betti numbers of the tangential variety $\mathcal{T}_g$.

2.3.1 Weyman Modules

We assume throughout this section that $p = \operatorname{char}(\mathbf{k}) \neq 2$. Let U be a $\mathbf{k}$-vector space of dimension two, and fix a basis $(1, x)$ so that $\operatorname{Sym}^d U$ can be identified with the space of polynomials of degree at most d in x. With this choice of basis we identify $\mathbf{k} \simeq \bigwedge^2 U$ via $1 \mapsto 1 \wedge x$. The perfect pairing

$$U \times U \longrightarrow \overset{2}{\bigwedge} U \simeq \mathbf{k}$$

gives rise to an identification $U \simeq U^\vee$ which we will use freely. For instance we will identify $(\operatorname{Sym}^d U)^\vee \simeq \operatorname{D}^d U$, instead of the more natural isomorphism $(\operatorname{Sym}^d U)^\vee \simeq \operatorname{D}^d(U^\vee)$ (see the Appendix). Recall that in characteristic zero (or larger than d) we also have an isomorphism $\operatorname{D}^d U \simeq \operatorname{Sym}^d U$, so that $\operatorname{Sym}^d U$ is isomorphic to its dual!

For $d \geq 0$ we consider the map

$$\psi : \bigwedge^2 \mathrm{Sym}^d U \longrightarrow \mathrm{Sym}^{2d-2} U, \tag{2.21}$$

$$\psi(x^i \wedge x^j) = (i-j) \cdot x^{i+j-1} \text{ for } 0 \leq i, j \leq d.$$

Exercise 2.11 Show that ψ is surjective (under our assumption that $\mathrm{char}(\mathbf{k}) \neq 2$).

Exercise 2.12 Check that the map ψ is $\mathrm{SL}(U)$-equivariant, where $\mathrm{SL}(U)$ is the group of linear automorphisms of U with determinant one. All the identifications that we make in this section will be $\mathrm{SL}(U)$-equivariant!

We define a pair (V, K) by setting $V^\vee = \mathrm{Sym}^d U$ and $K^\perp = \ker(\psi)$. It follows that $V = \mathrm{D}^d U$, and since ψ is surjective, we get

$$K^\vee = \bigwedge^2 V^\vee / K^\perp = \bigwedge^2 V^\vee / \ker(\psi) \simeq \mathrm{Im}(\psi) = \mathrm{Sym}^{2d-2} U,$$

so that $K = \mathrm{D}^{2d-2} U$. The Koszul module associated with this pair (V, K) is denoted

$$W^{(d)} := W(\mathrm{D}^d U, \mathrm{D}^{2d-2} U),$$

and is called a **Weyman module**. We next show that in most characteristics, Weyman modules satisfy the hypotheses of Theorems 2.2 and 2.3.

Lemma 2.2 *Let $p = \mathrm{char}(\mathbf{k})$. If $p = 0$ or $p \geq n$, then $W^{(n-1)}$ has vanishing resonance (and therefore it has finite length).*

Proof By (2.13), the vanishing of resonance is equivalent to the condition that $\ker(\psi)$ contains no non-zero decomposable form $f_1 \wedge f_2$. Suppose by contradiction that $\ker(\psi)$ contains a non-zero decomposable form $f_1 \wedge f_2$, and that $p = 0$ or $p \geq n$. By rescaling $f_1 \wedge f_2$, we may assume that f_1, f_2 are monic polynomials in x, say

$$f_1 = x^r + b_{r-1} x^{r-1} + \cdots + b_0, \quad f_2 = x^s + c_{s-1} x^{s-1} + \cdots + c_0.$$

If $r \neq s$, we may assume $r > s$ and we get $\psi(f_1 \wedge f_2) = (r-s) x^{r+s-1} + h(x)$, where $\deg(h) < r+s-1$ and $0 < r-s \leq n-1$. It follows that p cannot divide $r-s$, hence $\psi(f_1 \wedge f_2) \neq 0$, a contradiction. If $r = s$, then we can use the fact that $f_1 \wedge f_2 = f_1 \wedge (f_2 - f_1)$ to reduce to the case $r \neq s$. □

Exercise 2.13 Check that if $3 \leq p \leq n-1$ then $W^{(n-1)}$ has infinite length.

The relationship with the syzygies of $\mathcal{T}_g$ is given in the following theorem (see [2, Theorem 1.7] and Sect. 2.3.4).

Theorem 2.4 *If* char(**k**) $\neq$ 2, *then for each* $i = 1, \ldots, g - 3$, *we have a natural identification*

$$B_{i,2}(\mathcal{T}_g) = W^{(i+2)}_{g-3-i}.$$

Example 2.3 Let $g = 6$ and suppose that char(**k**) $= 0$. Combining Theorem 2.4 with (2.20) we get that the Betti table of $\mathcal{T}_6$ is

	0	1	2	3	4
0	1	–	–	–	–
1	–	6	5	–	–
2	–	–	5	6	–
3	–	–	–	–	1

Exercise 2.14 Using Theorems 2.3 and 2.4, write down the Betti table $\beta(\mathcal{T}_g)$ when char(**k**) $= 0$ and $g = 7, 8, 9$ (compare with Exercise 2.7).

Based on Theorem 2.4, we can now finish the proof of Theorem 2.1.

Proof of Theorem 2.1—The Equivalence (2.9) The implication "$\Longleftarrow$" in (2.9) was already discussed in Exercise 2.5. For the converse, we assume that $3 \leq p \leq \frac{g+1}{2}$, and using Theorem 2.4, we have to show that

$$W^{(i+2)}_{g-3-i} = 0 \text{ for } i \leq p - 3.$$

Let $n = i + 3$ and note that $p \geq n$, so Lemma 2.2 applies to show that $W^{(i+2)} = W^{(n-1)}$ has finite length. We also have that $p \geq n - 2$, so by Theorem 2.2 we conclude that $W^{(n-1)}_q = 0$ for $q \geq n - 3$. Since $g \geq 2p - 1$ and $i \leq p - 3$, it follows that if we let $q = g - 3 - i$ then $q \geq (2p - 1) - p = p - 1 \geq n - 3$. $\qquad\square$

Exercise 2.15 Prove the equivalence (2.8) using Theorem 2.4.

2.3.2 Hermite Reciprocity

Over the complex numbers, there are isomorphisms of $SL_2(\mathbb{C})$-representations

$$\text{Sym}^d(\text{Sym}^i\,\mathbb{C}^2) \simeq \text{Sym}^i(\text{Sym}^d\,\mathbb{C}^2) \simeq \bigwedge^i(\text{Sym}^{d+i-1}\,\mathbb{C}^2), \qquad (2.22)$$

according to [8, Exercises 11.34, 11.35], which we refer to as versions of **Hermite reciprocity**. Since representations of $SL_2(\mathbb{C})$ are completely reducible, the existence of such isomorphisms can be verified by a character calculation, that is, by identifying the torus weights and their multiplicities inside each of the three representations. A more interesting problem is to write down explicit isomorphisms between the representations in (2.22). Following [2, Section 3.4], we explain how this can be done, and also extended to an arbitrary field (where complete reducibility no longer holds, and certain symmetric power functors have to be replaced by divided powers).

Let U denote a 2-dimensional $\mathbf{k}$-vector space with basis $(1, x)$ as before. We will describe an explicit isomorphism

$$\Theta_d^i : \operatorname{Sym}^d (\mathrm{D}^i U) \longrightarrow \bigwedge^i (\operatorname{Sym}^{d+i-1} U). \tag{2.23}$$

Exercise 2.16

(a) Let V be an $SL(U)$-representation of dimension N. After choosing an isomorphism $\mathbf{k} \simeq \bigwedge^N V$, show that for each $r = 0, \cdots , N$ there exists isomorphisms

$$\bigwedge^r V \simeq \bigwedge^{N-r} V^\vee.$$

(b) Use part (a) and the existence of Θ_d^i to produce a chain of isomorphisms

$$\operatorname{Sym}^d (\mathrm{D}^i U) \cong \bigwedge^d (\operatorname{Sym}^{d+i-1} U)^\vee \cong (\operatorname{Sym}^i (\mathrm{D}^d U))^\vee$$

$$\cong \mathrm{D}^i (\operatorname{Sym}^d (U^\vee)) \cong \mathrm{D}^i (\operatorname{Sym}^d U).$$

The divided power $\mathrm{D}^i U$ has a basis consisting of elements $x^{(t)}, 0 \leq t \leq i$, from which we construct a basis for $\operatorname{Sym}^d (\mathrm{D}^i U)$ consisting of elements

$$e_\mu(x) = x^{(\mu_1)} \cdot x^{(\mu_2)} \cdots x^{(\mu_d)} \tag{2.24}$$

where $\mu = (\mu_1 \geq \cdots \geq \mu_d)$ is a partition with $\mu_1 \leq i$ and with at most d parts. Similarly, $\bigwedge^i (\operatorname{Sym}^{d+i-1} U)$ has a basis consisting of elements

$$s_\lambda(x) := x^{\lambda_1+i-1} \wedge x^{\lambda_2+i-2} \wedge \cdots \wedge x^{\lambda_i} \tag{2.25}$$

where $\lambda = (\lambda_1 \geq \cdots \geq \lambda_i)$ consists of partitions with $\lambda_1 \leq d$ and with at most i parts.

Consider the polynomial ring $\mathbf{k}[z_1, \cdots , z_i]$, and the subring consisting of symmetric polynomials. Examples of such are the **elementary symmetric polynomials**

$$e_r(z_1, \cdots , z_i) = \sum_{j_1 < \cdots < j_r} z_{j_1} \cdots z_{j_r},$$

which in turn can be used to define

$$e_\mu(z_1, \cdots, z_i) = e_{\mu_1}(z_1, \cdots, z_i) \cdots e_{\mu_d}(z_1, \cdots, z_i)$$

for every partition $\mu = (\mu_1 \geq \cdots \geq \mu_d)$. Other important examples are the **Schur polynomials**

$$s_\lambda(z_1, \ldots, z_i) = \frac{\det(z_k^{\lambda_\ell + i - k})_{1 \leq k, \ell \leq i}}{\det(z_k^{i-k})_{1 \leq k, \ell \leq i}},$$

defined for partitions $\lambda = (\lambda_1 \geq \cdots \geq \lambda_i)$. Using the basic theory of symmetric functions, particularly [11, Section I.6], one can solve the following.

Exercise 2.17 Show that the **k**-vector space with basis

$$\{e_\mu(z_1, \cdots, z_i) : i \geq \mu_1 \geq \cdots \geq \mu_d\}$$

coincides with the **k**-vector space with basis

$$\{s_\lambda(z_1, \cdots, z_i) : d \geq \lambda_1 \geq \cdots \geq \lambda_i\}.$$

If we let P_d^i denote the vector space of symmetric polynomials in Exercise 2.17 then the isomorphism Θ_d^i in (2.23) is constructed by composing the **k**-linear isomorphisms

$$\mathrm{Sym}^d(\mathrm{D}^i U) \longrightarrow P_d^i, \quad e_\mu(x) \mapsto e_\mu(z_1, \cdots, z_i)$$

and

$$P_d^i \xrightarrow{\ i\ } \bigwedge(\mathrm{Sym}^{d+i-1} U), \quad s_\lambda(z_1, \cdots, z_i) \mapsto s_\lambda(x).$$

Example 2.4 Let $i = d = 2$ (so that the partitions λ, μ vary over the same set). The following table records the corresponding symmetric polynomials.

λ	$(0,0)$	$(1,0)$	$(2,0)$	$(1,1)$	$(2,1)$	$(2,2)$
$e_\lambda(z_1, z_2)$	1	$z_1 + z_2$	$z_1 z_2$	$(z_1+z_2)^2$	$z_1 z_2(z_1+z_2)$	$(z_1 z_2)^2$
$s_\lambda(z_1, z_2)$	1	$z_1 + z_2$	$z_1^2 + z_1 z_2 + z_2^2$	$z_1 z_2$	$z_1 z_2(z_1+z_2)$	$(z_1 z_2)^2$

The change of basis is given by $e_{(2,0)} = s_{(1,1)}$, $e_{(1,1)} = s_{(2,0)} + s_{(1,1)}$, and $e_\lambda = s_\lambda$ for the remaining partitions. In particular, the map Θ_2^2 will satisfy

$$\Theta_2^2(x^{(2)} \cdot 1) = x^2 \wedge x \quad \text{and} \quad \Theta_2^2(x^{(1)} \cdot x^{(1)}) = x^3 \wedge 1 + x^2 \wedge x.$$

2.3.3 The Coordinate Independent Description of Γ_g

Let U denote a 2-dimensional **k**-vector space as before, and let $\mathbb{P}U \simeq \mathbf{P}^1$ denote the corresponding projective line. More precisely, if we let

$$A = \mathrm{Sym}(U) = \mathbf{k} \oplus U \oplus \mathrm{Sym}^2 U \oplus \cdots$$

denote the symmetric algebra of U, then $\mathbb{P}U = \mathrm{Proj}(A)$. If we choose a basis $\{x, y\}$ for U, then we get an identification $A \simeq \mathbf{k}[x, y]$, but we will avoid doing so in this section. The **Veronese subalgebra**

$$A^{(g)} = \bigoplus_{d \geq 0} \mathrm{Sym}^{dg} U,$$

is naturally a quotient of the symmetric algebra $S = \mathrm{Sym}(\mathrm{Sym}^g U)$, and this gives rise to a closed embedding

$$\Gamma_g \simeq \mathrm{Proj}(A^{(g)}) \hookrightarrow \mathrm{Proj}(S) \simeq \mathbf{P}^g.$$

The syzygy modules of Γ_g can be described in a coordinate-free fashion (using the Eagon–Northcott complex) by

$$B_{i,1}(\Gamma_g) = \mathrm{D}^{i-1} U \otimes \bigwedge^{i+1}(\mathrm{Sym}^{g-1} U) \text{ for } i = 1, \cdots, g - 1.$$

This description is more refined than the earlier calculation of Betti numbers, which can be recovered by noting that $\dim(\mathrm{D}^{i-1} U) = i$ and $\dim(\mathrm{Sym}^{g-1} U) = g$. What is even more remarkable is that if we fix i and vary g, then it can be shown using the explicit Hermite reciprocity of the previous section that

$$\mathfrak{B}_i = \bigoplus_{g \geq i+1} B_{i,1}(\Gamma_g)$$

has the structure of a free module over $\mathrm{Sym}(\mathrm{D}^i U)$, namely

$$\mathfrak{B}_i = \mathrm{D}^{i-1} U \otimes \mathrm{Sym}(\mathrm{D}^i U).$$

2.3.4 The Key Insight for the Proof of Theorem 2.4

The proof of Theorem 2.4 is quite involved (see [2, Section 5] for details), but the key idea is to realize that, just as in the case of the syzygies of the rational normal curve, if we fix i and let g vary, then

$$\bigoplus_{g \geq i+3} B_{i,2}(\mathcal{T}_g) \tag{2.26}$$

has the structure of a module over a polynomial ring, namely over $\text{Sym}(D^{i+2}U)$. This module is no longer free, and the main technical part of the work is to translate via Hermite reciprocity the calculation of $B_{i,2}(\mathcal{T}_g)$, and show that the resulting module from (2.26) is in fact the Weyman module $W^{(i+2)}$.

2.4 Syzygies of Curves

Let X be a non-singular projective curve of genus g, that is,

$$\dim H^0(X, \Omega_X) = g.$$

We define the **gonality** of X to be the smallest degree of a map $X \longrightarrow \mathbf{P}^1$, so that $X \simeq \mathbf{P}^1$ if and only if $\text{gon}(X) = 1$. If $\text{gon}(X) = 2$ then we say that X is **hyperelliptic**. If $g \geq 1$ and we fix a basis $\omega_1, \cdots, \omega_g$ of $H^0(X, \Omega_X)$ then we can define a morphism

$$\phi \colon X \longrightarrow \mathbf{P}^{g-1}, \ p \mapsto [\omega_1(p) : \cdots : \omega_g(p)],$$

and ϕ is a closed embedding if and only if X is not hyperelliptic. We call ϕ the **canonical map/embedding**. When X is hyperelliptic, ϕ maps X onto the rational normal curve Γ_{g-1} via a degree two map. When ϕ is an embedding, we can identify X with its image $\phi(X)$, which we refer to as a **canonical curve** (we note here that a canonical curve will have degree $2g - 2$). An important problem in the projective geometry of curves is the following (see [10, Section 5]):

Problem 2.3 Describe the defining equations, and the Betti tables of canonical curves.

Example 2.5 When $g = 3$, a canonical curve is simply a quartic curve in $\mathbf{P}^2$, so the Betti table is

	0	1
0	1	–
1	–	–
2	–	–
3	–	1

When $g = 4$, a canonical curve is a complete intersection of a quadric and a cubic surface in $\mathbf{P}^3$, with Betti table

	0	1	2
0	1	–	–
1	–	1	–
2	–	1	–
3	–	–	1

Theorem 2.5 (Petri [1, Section III.3]) *If X is general of genus $g \geq 5$ then the ideal $I(X)$ is generated by quadrics. The exceptions are the trigonal curves, and those isomorphic with plane quintics (which have genus 6).*

Example 2.6 When $g = 5$, the general canonical curves will have Betti table

	0	1	2	3
0	1	–	–	–
1	–	3	–	–
2	–	–	3	–
3	–	–	–	1

while for the trigonal ones the Betti table is

	0	1	2	3
0	1	–	–	–
1	–	3	2	–
2	–	2	3	–
3	–	–	–	1

Any canonical curve is Gorenstein, of Castelnuovo–Mumford regularity 3, and Green's Conjecture [10, Conjecture 5.1] predicts that the vanishing behavior of the Betti numbers $b_{i,j}(X)$ characterizes a geometric invariant of X called the **Clifford index** (for most curves, this is equal to $\mathrm{gon}(X) - 2$, and for our discussion it is safe to assume that this is the case—see for instance Example 2.6, where a general curve of genus 5 has gonality 4).

Conjecture 2.1 (Generic Green [10, Conjecture 5.6]) If X is a general curve over a field of characteristic zero then

$$b_{i,2}(X) = 0 \text{ if and only if } i \leq \frac{g-3}{2}. \tag{2.27}$$

This conjecture was proved by Voisin [17, 18], and work of Schreyer [16] shows that the natural extension to positive characteristic p may fail for p small (for instance when $p = 2$ and $g = 7$, or $p = 3$ and $g = 9$). The following conjecture was proposed recently by Eisenbud and Schreyer [7, Conjecture 0.1].

Conjecture 2.2 Generic Green's Conjecture holds in characteristic $p \geq \frac{g-1}{2}$.

The general strategy for proving Generic Green's conjecture is to produce an example of a curve X_0 that satisfies (2.27), and then use the semicontinuity of the Betti numbers, as follows [2, Theorem 1.2].

Theorem 2.6 *Generic Green's Conjecture holds over a field of characteristic $p \geq \frac{g+2}{2}$.*

Proof *(Sketch)* Let X_0 denote a general hyperplane section of $\mathcal{T}_g$, so that

$$b_{i,j}(X_0) = b_{i,j}(\mathcal{T}_g) \text{ for all } i, j. \tag{2.28}$$

X_0 is a rational g-cuspidal curve, so it has (arithmetic) genus g. It can be realized as a limit of smooth curves of genus g, $(X_t)_t \longrightarrow X_0$. Using (2.28) and Theorem 2.1, it follows that

$$b_{i,2}(X_0) = 0 \text{ for } i \leq \frac{g-3}{2}. \tag{2.29}$$

Since Betti numbers are upper semicontinuous in families, it follows that $b_{i,2}(X_t) = 0$ for some (a general) t, so there is at least one smooth curve X_t satisfying the Generic Green Conjecture. Again by semicontinuity, the same is true for a curve X in a neighborhood of X_t in the (irreducible) moduli space of genus g curves. □

Using Theorem 2.1 and (2.28), it follows that the vanishing (2.29) fails when $p \leq \frac{g+1}{2}$, so the surface $\mathcal{T}_g$ can't be used to prove Conjecture 2.2. Nevertheless, by building a parallel theory of **bi-graded Koszul modules**, and replacing $\mathcal{T}_g$ with a **K3 carpet**, in [14] it is shown that Conjecture 2.2 holds in fact for $p \geq \left\lfloor \frac{g-1}{2} \right\rfloor$.

Appendix

In this Appendix we record some basic facts about complexes and multilinear algebra that are used in the chapter. For more details see [5, Appendix 2].

Symmetric vs Divided Powers

Let V denote a free module of finite rank over a ring $\mathbf{k}$, and for $d > 0$ consider the tensor power $T^d V := V^{\otimes d} = V \otimes V \otimes \cdots \otimes V$ with the natural action of the symmetric group $\mathfrak{S}_d$ by permuting the factors. The **divided power** $D^d V$ is defined as the set of symmetric tensors in $T^d V$, that is,

$$D^d V := \{\omega \in T^d V : \sigma(\omega) = \omega \text{ for all } \sigma \in \mathfrak{S}_d\}.$$

If we consider the subspace of $T^d V$ defined by

$$\Sigma_d := \text{Span}\{\sigma(\omega) - \omega : \sigma \in \mathfrak{S}_d \text{ and } \omega \in T^d V\},$$

then the **symmetric power** $\text{Sym}^d V$ is defined as the quotient $\text{Sym}^d V :=$ $T^d V/\Sigma_d$. If V has a basis $(x_1, \ldots, x_n)$, then $\text{Sym}^d V$ identifies with the space of homogeneous polynomials of degree d in $x_1, \ldots, x_n$, and as such it has a basis of monomials $x_1^{a_1} \cdots x_n^{a_n}$, where $a_1 + \cdots + a_n = d$. To get a basis for $D^d V$ we first consider the orbits

$$O_{a_1, \ldots, a_n} := \mathfrak{S}_d \cdot x_1^{\otimes a_1} \otimes x_2^{\otimes a_2} \otimes \cdots \otimes x_n^{\otimes a_n}$$

and for $a_1 + \cdots + a_n = d$ consider the **divided power monomials**

$$x_1^{(a_1)} \cdots x_n^{(a_n)} := \sum_{\omega \in O_{a_1, \ldots, a_n}} \omega.$$

They form a basis for $D^d(V)$, and in particular we have that $\dim(\text{Sym}^d V) = \dim(D^d V)$. By composing the inclusion of $D^d V$ into $T^d V$ with the projection onto $\text{Sym}^d V$ we obtain a natural map

$$D^d V \longrightarrow \text{Sym}^d V, \quad x_1^{(a_1)} \cdots x_n^{(a_n)} \mapsto \binom{d}{a_1, \ldots, a_n} x_1^{a_1} \cdots x_n^{a_n},$$

$$\text{where} \quad \binom{d}{a_1, \ldots, a_n} = \frac{d!}{a_1! \cdots a_n!}. \tag{2.30}$$

This map is an isomorphism when the multinomial coefficients are invertible (for instance, when **k** is a field with char(**k**) = 0 or char(**k**) > d). In general it is neither injective, nor surjective. There is a natural $\mathfrak{S}_n$-invariant perfect pairing

$$T^d(V) \times T^d(V^\vee) \longrightarrow \mathbf{k}, \tag{2.31}$$

defined on pure tensors via

$$\langle v_1 \otimes \cdots \otimes v_d, f_1 \otimes \cdots \otimes f_d \rangle = f_1(v_1) f_2(v_2) \cdots f_d(v_d), \text{ where } v_i \in V \text{ and } f_j \in V^\vee.$$

This induces a perfect pairing between $\text{Sym}^d V$ and $D^d(V^\vee)$, giving a natural identification

$$(\text{Sym}^d V)^\vee \cong D^d(V^\vee).$$

The construction of tensor, symmetric and divided powers is functorial, so it can be applied to any locally free sheaf $\mathcal{E}$ on a variety X. It follows from the discussion above that when $\operatorname{char}(\mathbf{k}) > d$ we have $\operatorname{Sym}^d(\mathcal{E}) \simeq D^d(\mathcal{E})$, and the inclusion $D^d(\mathcal{E}) \hookrightarrow T^d(\mathcal{E})$ is split. In arbitrary characteristic, we have $(\operatorname{Sym}^d \mathcal{E})^\vee \simeq D^d(\mathcal{E}^\vee)$, where $\mathcal{E}^\vee = \mathcal{H}om_{O_X}(\mathcal{E}, O_X)$.

The Eagon–Northcott and Koszul Complexes

Suppose that F, G are free modules over a ring R, with $\operatorname{rank}(F) = n$, $\operatorname{rank}(G) = m$, $n \geq m$, and consider an R-linear map $\alpha \colon F \longrightarrow G$. The **Eagon–Northcott complex** of the map α is denoted $\mathbf{EN}_\bullet(\alpha)$, and its terms are given by

$$\mathbf{EN}_0(\alpha) = \overset{m}{\bigwedge} G, \quad \mathbf{EN}_i(\alpha) = \overset{i+m-1}{\bigwedge} F \otimes D^{i-1}(G^\vee) \text{ for } i = 1, \cdots, n-m+1.$$

The differential $d_i \colon \mathbf{EN}_i \longrightarrow \mathbf{EN}_{i-1}$ is constructed as follows. The first map is $d_1 = \bigwedge^m \alpha$, and for $i \geq 2$, d_i is given by

$$\overset{i+m-1}{\bigwedge} F \otimes D^{i-1}(G^\vee) \longrightarrow \left(\overset{i+m-2}{\bigwedge} F \otimes F \right) \otimes (D^{i-2}(G^\vee) \otimes G^\vee) \longrightarrow \overset{i+m-2}{\bigwedge} F \otimes D^{i-2}(G^\vee),$$

where the first map is induced by the natural inclusions, and the second one is induced by

$$\alpha \in \operatorname{Hom}_R(F, G) = F^\vee \otimes G = \operatorname{Hom}_R(F \otimes G^\vee, R).$$

If α is surjective then $\mathbf{EN}_\bullet(\alpha)$ is an exact complex.

If we choose bases for F and G, then α is expressed by a $m \times n$ matrix, and $\bigwedge^m \alpha$ is given by a one-row matrix whose entries are the $m \times m$ minors of α. Writing $I_m(\alpha)$ for the ideal they generate, it follows that $\operatorname{coker}(d_1) \simeq S/I_m(\alpha)$. Suppose that R is a standard graded polynomial ring, that F, G are graded R-modules, and that α is a degree preserving map and it is minimal (that is, it has entries in the maximal homogeneous ideal). Under suitable genericity assumptions, such as $\operatorname{codim}(I_m(\alpha)) = n - m + 1$, $\mathbf{EN}_\bullet(\alpha)$ gives a minimal resolution of $S/I_m(\alpha)$.

The complex obtained in the special case $m = 1$ is called a **Koszul complex**, and is denoted $\mathbf{K}_\bullet(\alpha)$. The condition that $\operatorname{codim}(I_m(\alpha)) = n - m + 1$ is then equivalent to the requirement that the entries of α form a regular sequence, which is the familiar condition characterizing the exactness of the Koszul complex.

By functoriality, one can perform similar constructions in the relative setting, when F, G are replaced by locally free sheaves $\mathcal{F}, \mathcal{G}$ on a variety X. If α is surjective then $\mathbf{EN}_\bullet(\alpha)$ is an exact complex, as remarked in the absolute setting.

The examples of interest for us are those coming from a rational normal curve, respectively from a rational normal scroll. If $R = S = k[z_0, \cdots, z_g]$, $m = 2$, $n = g$ and α is given by the matrix Z in (2.2), then $n - m + 1 = g - 1$ is the codimension of Γ_g in $\mathbf{P}^g$, and the corresponding Eagon–Northcott complex gives a minimal resolution of $S/I_2(\alpha)$.

Exercise 2.18 Choose bases and write down explicitly the Eagon–Northcott complex for the rational normal curves of degree $g = 3$ and $g = 4$.

For $a_1, \cdots, a_d \geq 1$, a **rational normal scroll** of type $(a_1, a_2, \cdots, a_d)$ is defined by the 2×2 minors of a matrix obtained by concatenating matrices of the form (2.2) in different sets of variables:

$$A(a_1, \cdots, a_d) = \begin{pmatrix} x_{1,0} \ x_{1,1} \ \cdots \ x_{1,a_1-1} & \cdots \cdots & x_{d,0} \ x_{d,1} \ \cdots \ x_{d,a_d-1} \\ x_{1,1} \ x_{1,2} \ \cdots \ x_{1,a_1} & \cdots \cdots & x_{d,1} \ x_{d,2} \ \cdots \ x_{d,a_d} \end{pmatrix}$$

The resulting scroll has dimension d and is a subvariety in $\mathbf{P}^{a_1+\cdots+a_d+d-1}$. Note that the matrix $A(a_1, \cdots, a_d)$ gives a map between free modules of ranks $n = a_1 + \cdots + a_d$ and $m = 2$, so $n - m + 1$ is precisely the codimension of the scroll, and the Eagon–Northcott complex is again exact. The case of the rational normal curve of degree g can be recovered by taking $d = 1$ and $a_1 = g$.

Exercise 2.19 Verify that the columns of the matrices (2.6) (resp. (2.10)) can be rearranged in such a way that they agree, after a relabelling, with scrollar matrices $A(a_1, a_2)$ (resp. $A(a_1, \cdots, a_p)$).

The Buchsbaum–Rim Complex

We continue with the notation from the previous section. The **Buchsbaum–Rim** complex associated with the map $\alpha : F \longrightarrow G$ is denoted $\mathbf{BR}_\bullet(\alpha)$, and its terms are given by

$$\mathbf{BR}_0(\alpha) = G, \quad \mathbf{BR}_1(\alpha) = F,$$

$$\mathbf{BR}_i(\alpha) = \bigwedge^{i+m-1} F \otimes \det(G^\vee) \otimes D^{i-2}(G^\vee) \text{ for } i = 2, \cdots, n - m + 1.$$

The first differential is $d_1 = \alpha$, while the second one is obtained by composing

$$\bigwedge^{m+1} F \otimes \bigwedge^m G^\vee \longrightarrow \left(\bigwedge^m F \otimes F \right) \otimes \bigwedge^m G^\vee \longrightarrow F$$

where the first map is the natural inclusion, while the second one is induced by

$$\bigwedge^m \alpha \in \text{Hom}_R \left(\bigwedge^m F, \bigwedge^m G \right) = \text{Hom}_R \left(\bigwedge^m F \otimes \bigwedge^m G^\vee, R \right).$$

The higher differentials are defined in analogy with the ones for the Eagon–Northcott complex. Just like in the case of the Eagon–Northcott complex, when α is surjective the complex $\mathbf{BR}_\bullet(\alpha)$ is exact. A similar statement holds if we replace F, G by locally free sheaves.

References

1. E. Arbarello, M. Cornalba, P. A. Griffiths, J. Harris, *Geometry of Algebraic Curves*, vol. I. Grundlehren der Mathematischen Wissenschaften [Fundamental Principles of Mathematical Sciences], vol. 267 (Springer, New York, 1985)
2. M. Aprodu, G. Farkas, Ş. Papadima, C. Raicu, J. Weyman, Koszul modules and Green's conjecture. Invent. Math. **218**(3), 657–720 (2019)
3. D. Cox, J. Little, D. O'Shea, *Ideals, Varieties, and Algorithms. An Introduction to Computational Algebraic Geometry and Commutative Algebra*, 3rd edn. Undergraduate Texts in Mathematics (Springer, New York, 2007)
4. D. Eisenbud, *Green's Conjecture: An Orientation for Algebraists. Free Resolutions in Commutative Algebra and Algebraic Geometry (Sundance, UT, 1990)*. Res. Notes Math., vol. 2 (Jones and Bartlett, Boston, MA, 1992), pp. 51–78
5. D. Eisenbud, *Commutative Algebra, with a View Toward Algebraic Geometry*. Graduate Texts in Mathematics, vol. 150 (Springer, New York, 1995)
6. D. Eisenbud, *The Geometry of Syzygies. A Second Course in Commutative Algebra and Algebraic Geometry*. Graduate Texts in Mathematics, vol. 229 (Springer, New York, 2005)
7. D. Eisenbud, F.-O. Schreyer, Equations and syzygies of K3 carpets and unions of scrolls. Acta Math. Vietnam. **44**(1), 3–29 (2019)
8. W. Fulton, J. Harris, *Representation Theory. A Girst Course*. Graduate Texts in Mathematics, vol. 129 (Springer, New York, 1991)
9. D.R. Grayson, M.E. Stillman, *Macaulay 2, A Software System for Research in Algebraic Geometry.*
10. M. Green, Koszul cohomology and the geometry of projective verieties. J. Differ. Geom. **19**, 125–171 (1984)
11. I.G. Macdonald, Symmetric functions and Hall polynomials. Oxford Math. Monographs (1995)
12. P.J. Olver, *Classical Invariant Theory*. London Mathematical Society Student Texts, vol. 44 (Cambridge University Press, Cambridge, 1999)
13. Ş. Papadima, A.I. Suciu, Vanishing resonance and representations of Lie algebras. J. Reine Angew. Math. **706**, 83–101 (2015)
14. C. Raicu, S.V. Sam, *Bi-graded Koszul modules, K3 carpets, and Green's conjecture.* arXiv:1909.09122
15. F.-O. Schreyer, Syzygies of curves with special pencils. Thesis (Ph.D.)–Brandeis University, 1983
16. F.-O. Schreyer, Syzygies of canonical curves and special linear series. Mathematische Annalen **275**, 105–137 (1986)
17. C. Voisin, Green's generic syzygy conjecture for curves of even genus lying on a K3 surface. J. Euro. Math. Soc. **4**, 363–404 (2002)
18. C. Voisin, Green's canonical syzygy conjecture for generic curves of odd genus. Compositio Math. **141**, 1163–1190 (2005)

Chapter 3
Gröbner Degenerations

Matteo Varbaro

Abstract We will review the basics of Gröbner basis theory, with a special focus on the deformation aspect. After this, we will give a self-contained proof of the recent result achieved in (Conca and Varbaro, Invent. math. **221**(3), 713–730 (2020)) that, if a homogeneous polynomial ideal admits a squarefree monomial initial ideal, then the local cohomology modules of the original ideal and those of such initial ideal have the same Hilbert functions.

3.1 Gröbner Bases for Beginners

In this section we will review the basics of Gröbner basis theory, up to a description of the Buchberger algorithm. What we develop here is just enough for our purposes, for a more extensive exposition see the chapter on Gröbner bases of Eisenbud's book [8].

3.1.1 Notation and Basic Definitions

Throughout this chapter the symbol $\mathbb{N}$ denotes the set $\{0, 1, 2, \ldots\}$; we fix a base field K and the polynomial ring $R = K[X_1, \ldots, X_n]$ in n variables over K. We provide R with a graded structure by putting $\deg(X_i) = g_i$, where $g = (g_1, \ldots, g_n)$ is a vector of positive integers. Note that in this way $\mathfrak{m} := (X_1, \ldots, X_n)$ is the unique homogeneous maximal ideal of R. When we speak of homogeneous ideals of R or of graded R-modules we mean with respect to such grading: if we want to consider the standard grading in which all variables have degree 1 we will say it explicitly.

M. Varbaro (✉)
Università di Genova, DIMA, Genoa, Italy
e-mail: varbaro@dima.unige.it

© The Editor(s) (if applicable) and The Author(s), under exclusive license
to Springer Nature Switzerland AG 2021
A. Conca et al. (eds.), *Recent Developments in Commutative Algebra*, Lecture
Notes in Mathematics 2283, https://doi.org/10.1007/978-3-030-65064-3_3

A *monomial* of R is an element $\mathbf{X^u} := X_1^{u_1} \cdots X_n^{u_n} \in R$, where $\mathbf{u} = (u_1, \ldots, u_n) \in \mathbb{N}^n$, and the set of monomials of R is denoted by $\mathrm{Mon}(R)$. A *term* of R is an element of the form $a\mu \in R$ where $a \in K$ and μ is a monomial. Notice that every $f \in R$ can be written as a sum of terms: there exists a unique (finite) subset $\mathrm{supp}(f) \subset \mathrm{Mon}(R)$ such that

$$f = \sum_{\mu \in \mathrm{supp}(f)} a_\mu \mu, \quad a_\mu \in K \setminus \{0\}.$$

In the above representation, the only lack of uniqueness amounts to the order of the terms.

Definition 3.1 A *monomial order* on R is a total order $<$ on $\mathrm{Mon}(R)$ such that:

1. $1 \le \mu$ for every $\mu \in \mathrm{Mon}(R)$;
2. If $\mu_1, \mu_2, \nu \in \mathrm{Mon}(R)$ are such that $\mu_1 \le \mu_2$, then $\mu_1 \nu \le \mu_2 \nu$.

Remark 3.1 If $<$ is a monomial order on R and μ, ν are monomials such that $\mu | \nu$, then $\mu \le \nu$: indeed $1 \le \nu/\mu$, so $\mu = 1 \cdot \mu \le (\nu/\mu) \cdot \mu = \nu$. Therefore, a monomial order on R is a total order extending the partial order on $\mathrm{Mon}(R)$ given by divisibility. On the other hand, as soon as $n \ge 2$, it is easy to find total orders extending the divisibility order which fail to be monomial orders. We will have several opportunities to convince ourselves that the more restrictive notion of monomial order is necessary to develop the theory of Gröbner bases.

Example 3.1 Typical examples of monomial orders are the following: given two monomials $\mu = X_1^{u_1} \cdots X_n^{u_n}$ and $\nu = X_1^{v_1} \cdots X_n^{v_n}$ we define

- The *lexicographic order* (Lex) by $\mu <_{\mathrm{Lex}} \nu$ if $u_k < v_k$ for some k and $u_i = v_i$ for any $i < k$.
- The *degree lexicographic order* (DegLex) by $\mu <_{\mathrm{DegLex}} \nu$ if $\deg(\mu) < \deg(\nu)$ or $\deg(\mu) = \deg(\nu)$ and $\mu <_{\mathrm{Lex}} \nu$.
- The *(degree) reverse lexicographic order* (RevLex) by $\mu <_{\mathrm{RevLex}} \nu$ if $\deg(\mu) < \deg(\nu)$ or $\deg(\mu) = \deg(\nu)$ and $u_k > v_k$ for some k and $u_i = v_i$ for any $i > k$.

Example 3.2 In $K[X, Y, Z]$, assuming that $X > Y > Z$ and the grading is standard, we have $X^2 >_{\mathrm{Lex}} XZ >_{\mathrm{Lex}} Y^2$, while $X^2 >_{\mathrm{RevLex}} Y^2 >_{\mathrm{RevLex}} XZ$.

The fact that monomial orders extend the divisibility order implies the following important property:

Proposition 3.1 *A monomial order on $R = K[X_1, \ldots, X_n]$ is a well-order on $\mathrm{Mon}(R)$; that is, any nonempty subset of $\mathrm{Mon}(R)$ has a minimum. Equivalently, all descending chains of monomials in R terminate.*

Proof Let $\emptyset \ne N \subset \mathrm{Mon}(R)$, and $I \subset R$ be the ideal generated by N. By the Hilbert basis theorem, I is generated by a finite number of monomials of N. Since a monomial order refines divisibility, the minimum of such finitely many monomials is also the minimum of N. $\square$

3.1.2 Gröbner Bases

From now on, we fix a monomial order $<$ on $R = K[X_1, \ldots, X_n]$, so that every polynomial $0 \neq f \in R$ can be written uniquely as

$$f = a_1 \mu_1 + \ldots + a_k \mu_k$$

with $a_i \in K \setminus \{0\}$, $\mu_i \in \mathrm{Mon}(R)$ and $\mu_1 > \mu_2 > \ldots > \mu_k$.

Definition 3.2 The *initial monomial* of f is $\mathrm{in}_<(f) = \mu_1$. Furthermore, its *initial coefficient* is $\mathrm{inic}_<(f) = a_1$ and its *initial term* is $\mathrm{init}_<(f) = a_1 \mu_1$.

Remark 3.2 When no confusion can arise, we will suppress the subscript "$<$" in the above definition.

For $f, g \in R \setminus \{0\}$ we have the following relations:

- $\mathrm{inic}(f)\mathrm{in}(f) = \mathrm{init}(f)$.
- $\mathrm{in}(fg) = \mathrm{in}(f)\mathrm{in}(g)$.
- $\mathrm{in}(f + g) \leq \max\{\mathrm{in}(f), \mathrm{in}(g)\}$.

Example 3.3 If $f = X_1 + X_2 X_4 + X_3^2$, assuming that the grading is standard we have:

- $\mathrm{in}(f) = X_1$ with respect to Lex.
- $\mathrm{in}(f) = X_2 X_4$ with respect to DegLex.
- $\mathrm{in}(f) = X_3^2$ with respect to RevLex.

Example 3.4 If $f = X^2 + XY + Y^2 \in K[X, Y]$ then, for any monomial order, we have:

- $\mathrm{in}(f) = X^2$ if $X > Y$.
- $\mathrm{in}(f) = Y^2$ if $Y > X$.

In particular, $XY \neq \mathrm{in}(f)$ whatever the monomial order is.

What happened in the above example can be seen more clearly via of the Newton polytope:

Definition 3.3 If $0 \neq f \in R = K[X_1, \ldots, X_n]$, the *Newton polytope* of f is the convex hull of $\{\mathbf{u} : \mathbf{X}^{\mathbf{u}} \in \mathrm{supp}(f)\} \subset \mathbb{R}^n$, and is denoted by $\mathrm{NP}(f) \subset \mathbb{R}^n$.

We will see later that, for a given nonzero polynomial $f \in R$, the exponent vector of $\mathrm{in}(f)$ must be a vertex of $\mathrm{NP}(f)$. In Example 3.4 we have $\mathrm{NP}(f) = \{(x, y) \in \mathbb{R}^2 : x + y = 2, x \geq 0\} \subset \mathbb{R}^2$, and obviously the exponent vector of XY, namely $(1, 1)$, is not a vertex of $\mathrm{NP}(f)$.

Definition 3.4 If I is an ideal of $R = K[X_1, \ldots, X_n]$, then the monomial ideal $in_<(I) \subset R$ generated by $\{in_<(f) : f \in I\} \subset R$ is named the *initial ideal* of I.

Remark 3.3 When no confusion can arise, we will suppress the subscript "$<$" in the above definition.

In the above definition notice that the K-vector space generated by $\{in(f) : f \in I\} \subset R$ is already an ideal of R, so one may also define $in(I) \subset R$ as the K-vector space generated by $\{in(f) : f \in I\} \subset R$.

Definition 3.5 Polynomials $f_1, \ldots, f_m$ of an ideal $I \subset R = K[X_1, \ldots, X_n]$ form a *Gröbner basis* of I if $in(I) = (in(f_1), \ldots, in(f_m))$.

Example 3.5 Consider the ideal $I = (f_1, f_2)$ of $K[X, Y, Z]$, where $f_1 = X^2 - Y^2$ and $f_2 = XZ - Y^2$, and assume that the grading is standard. For Lex with $X > Y > Z$ the polynomials f_1, f_2 are not a Gröbner basis of I, since $XY^2 = in(Zf_1 - Xf_2)$ is a monomial of $in(I)$ which is not in $(in(f_1), in(f_2)) = (X^2, XZ)$. For RevLex with $X > Y > Z$, one can show that $in(I) = (X^2, Y^2)$, so f_1 and f_2 are a Gröbner basis of I in this case.

Remark 3.4 The Noetherianity of R implies that any ideal of R has a finite Gröbner basis.

3.1.3 The Buchberger Algorithm

There is a way to compute a Gröbner basis of an ideal I starting from a system of generators of I, namely the *Buchberger algorithm*; such algorithm also checks if the given system of generators is already a Gröbner basis. To describe the Buchberger algorithm we need to introduce the following concept:

Definition 3.6 Let $f_1, \ldots, f_m \in R = K[X_1, \ldots, X_n]$. A polynomial $r \in R$ is a *reduction of $g \in R$ modulo $f_1, \ldots, f_m$* if there exist $q_1, \ldots, q_m \in R$ satisfying:

- $g = q_1 f_1 + \ldots + q_m f_m + r$;
- If $q_i \neq 0$, $in(q_i f_i) \leq in(g)$ for all $i = 1, \ldots, m$;
- For all $i = 1, \ldots, m$, $in(f_i)$ does not divide μ $\forall$ $\mu \in \text{supp}(r)$.

Lemma 3.1 Let $f_1, \ldots, f_m \in R = K[X_1, \ldots, X_n]$. Every polynomial $g \in R$ admits a reduction modulo $f_1, \ldots, f_m$.

Proof Let $J = (in(f_1), \ldots, in(f_m))$. We start with $r = g$ and apply the *reduction algorithm*:

1. If $\text{supp}(r) \cap J = \emptyset$, we are done: r is the desired reduction.
2. Otherwise choose $\mu \in \text{supp}(r) \cap J$ and let $b \in K$ be the coefficient of μ in r. Choose i such that $in(f_i) \mid \mu$ and set $r' = r - a v f_i$ where $v = \mu/in(f_i)$ and $a = b/\text{inic}(f_i)$. Then replace r by r' and go to 1.

This algorithm terminates after finitely many steps since it replaces the monomial μ by a linear combination of monomials that are smaller in the monomial order, and all descending chains of monomials in R terminate. We obtain the "quotients" q_i by accumulating the terms avf_i that arise in step 2. It is easy to check that they satisfy the conditions of Definition 3.6. □

Example 3.6 Once again, we take $R = K[X, Y, Z]$, $f_1 = X^2 - Y^2$ and $f_2 = XZ - Y^2$, and we consider Lex with $X > Y > Z$. Set $g = X^2 Z$. Then $g = Zf_1 + Y^2 Z$, but $g = Xf_2 + XY^2$ as well. Both these equations yield reductions of g, namely XY^2 and $Y^2 Z$. Thus a polynomial can have several reductions modulo f_1, f_2.

As it turns out, the reduction of $g \in R$ modulo $f_1, \ldots, f_m$ is unique when $f_1, \ldots, f_m$ is a Gröbner basis.

Proposition 3.2 *Let I be an ideal of $R = K[X_1, \ldots, X_n]$, $f_1, \ldots, f_m \in I$ and $J = (\mathrm{in}(f_1), \ldots, \mathrm{in}(f_m))$. Then the following are equivalent:*

1. *$f_1, \ldots, f_m$ form a Gröbner basis of I;*
2. *every $g \in I$ reduces to 0 modulo $f_1, \ldots, f_m$;*
3. *the monomials not in J are linearly independent modulo I.*

If the equivalent conditions above hold, then:

4. *Every element of R has a unique reduction modulo $f_1, \ldots, f_m$. Furthermore, such a reduction depends only on I and the monomial order, but not on the chosen Gröbner basis $f_1, \ldots, f_m$ of I.*

Proof

$1 \Rightarrow 3$ Assume that 3 is false. Then there exists a polynomial $r \in I$ which is a nonzero linear combination of monomials $\mu \notin J$. On the other hand, $\mathrm{in}(r)$ is divisible by $\mathrm{in}(f_i)$ for some i since $f_1, \ldots, f_m$ are a Gröbner basis. This is a contradiction.

$3 \Rightarrow 2$ Let r be a reduction of $g \in I$ modulo $f_1, \ldots, f_m$. Then $r \in I$ as well, so r vanishes modulo I. Being a linear combination of monomials $\mu \notin J$, which are linearly independent modulo I by assumption, it must be $r = 0$.

$2 \Rightarrow 1$ Let $g \in I$, $g \neq 0$. If g reduces to 0, then we have

$$g = q_1 f_1 + \cdots + q_m f_m$$

such that, if $q_i \neq 0$, $\mathrm{in}(q_i f_i) \leq \mathrm{in}(g)$ for all i. But the monomial $\mathrm{in}(g)$ must appear on the right hand side as well, and this is only possible if $\mathrm{in}(g) = \mathrm{in}(q_i f_i) = \mathrm{in}(q_i) \mathrm{in}(f_i)$ for at least one i. In other words, $\mathrm{in}(g)$ must be divisible by $\mathrm{in}(f_i)$ for some i. Hence $\mathrm{in}(I) = J$.

$3 \Rightarrow 4$ Let r_1, r_2 be reductions of $g \in R$. Then, on the one hand, $r_1 - r_2 \in I$, and, on the other hand, $r_1 - r_2$ is a linear combination of monomials $\mu \notin J$. By 3 we must have $r_1 = r_2$. Finally observe that, since $J = \mathrm{in}(I)$, the proof of uniqueness above does not depend on $f_1, \ldots, f_m$. □

As a corollary we get that a Gröbner basis of an ideal $I \subset R$ generates I. The following two further corollaries will be useful later.

Corollary 3.1 *Let $I \subset R = K[X_1, \ldots, X_n]$ be an ideal and $<_1, <_2$ monomial orders of R. If $\text{in}_{<_1}(I) \subset \text{in}_{<_2}(I)$, then $\text{in}_{<_1}(I) = \text{in}_{<_2}(I)$.*

Proof By Proposition 3.2, the sets A_i of monomials of R not in $\text{in}_{<_i}(I)$ is a K-bases of R/I for each $i = 1, 2$. Since $A_1 \supset A_2$, we must have $A_1 = A_2$. □

Corollary 3.2 *Let $I_1, I_2 \subset R = K[X_1, \ldots, X_n]$ be ideals and $<$ a monomial order of R. If $I_1 \subset I_2$ and $\text{in}_<(I_1) = \text{in}_<(I_2)$, then $I_1 = I_2$.*

Proof By Proposition 3.2, the set A of monomials of R not in $\text{in}_<(I_1) = \text{in}_<(I_2)$ are K-bases of R/I_i for each $i = 1, 2$. Since $I_1 \subset I_2$, we must have $I_1 = I_2$. □

Definition 3.7 The *S-polynomial* of two polynomials $f, g \in R$ is defined as

$$S(f, g) = \frac{\text{lcm}(\text{in}(f), \text{in}(g))}{\text{init}(f)} f - \frac{\text{lcm}(\text{in}(f), \text{in}(g))}{\text{init}(g)} g$$

We are ready to show the Buchberger algorithm.

Proposition 3.3 *Let $f_1, \ldots, f_m \in R = K[X_1, \ldots, X_n]$ and $I = (f_1, \ldots, f_m)$. Then the following are equivalent:*

1. *$f_1, \ldots, f_m$ form a Gröbner basis of I.*
2. *For all $1 \leq i < j \leq m$, $S(f_i, f_j)$ reduces to 0 modulo $f_1, \ldots, f_m$.*

Proof

$1 \Rightarrow 2$ It follows by Proposition 3.2 since $S(f_i, f_j) \in I$.
$2 \Rightarrow 1$ We need to show that every $g \in I$ reduces to 0 modulo the f_k's. Since $g \in I$, we have

$$g = a_1 f_1 + \ldots + a_m f_m$$

for some $a_k \in R$. Among such representations, we can choose one minimizing $\mu := \max\{\text{in}(a_i f_i) : i = 1, \ldots, m\}$ and, among these, one minimizing the cardinality of the set $\{i = 1, \ldots, m | \text{in}(a_i f_i) = \mu\}$; let us denote such a minimum by s. By contradiction, suppose that $\mu > \text{in}(g)$. In this case $s \geq 2$, so there exist $i < j$ such that $\text{in}(a_i f_i) = \text{in}(a_j f_j) = \mu$. Set $c := \text{inic}(a_i f_i)$ and notice that $\mu = v \cdot \alpha_{ij}$, where $\alpha_{ij} := \text{lcm}(\text{in}(f_i), \text{in}(f_j))$, for some $v \in \text{Mon}(R)$. Let

$$S(f_i, f_j) = q_1 f_1 + \ldots + q_m f_m$$

the reduction of $S(f_i, f_j)$ (so that, for all k, $\mathrm{in}(q_k f_k) \leq \mathrm{in}(S(f_i, f_j)) < \alpha_{ij}$). From this we get a representation $g = a'_1 f_1 + \ldots + a'_m f_m$ where

- $a'_i = a_i - \dfrac{cv\alpha_{ij}}{\mathrm{init}(f_i)} + cvq_i;$
- $a'_j = a_j + \dfrac{cv\alpha_{ij}}{\mathrm{init}(f_j)} + cvq_j;$
- $a'_k = a_k + cvq_k$ for $i \neq k \neq j$,

contradicting the minimality of μ and s.

$\square$

3.2 Initial Ideals with Respect to Weights

Fix $w = (w_1, \ldots, w_n) \in \mathbb{N}^n$ a *weight vector*. If $\mu = \mathbf{X}^{\mathbf{u}} \in \mathrm{Mon}(R)$ with $\mathbf{u} = (u_1, \ldots, u_n)$ then we set

$$w(\mu) := w_1 u_1 + \ldots + w_n u_n.$$

If $0 \neq f \in R = K[X_1, \ldots, X_n]$ we set $w(f) := \max\{w(\mu) : \mu \in \mathrm{supp}(f)\}$ and

$$\mathrm{init}_w(f) = \sum_{w(\mu)=w(f)} a_\mu \mu,$$

where $f = \sum_{\mu \in \mathrm{supp}(f)} a_\mu \mu$ (of course we are putting $a_\mu = 0$ if $\mu \notin \mathrm{supp}(f)$).

Example 3.7 If $w = (2, 1)$ and $f = X^3 + 2X^2Y^2 - Y^5 \in \mathbb{Q}[X, Y]$ then $\mathrm{init}_w(f) = X^3 + 2X^2Y^2$.

Given an ideal $I \subset R$ we set $\mathrm{in}_w(I) \subset R$ to be the ideal generated by $\{\mathrm{init}_w(f) : f \in I\}$. Since $\mathrm{in}_w(f)\mathrm{in}_w(g) = \mathrm{in}_w(fg)$ for any $f, g \in R \setminus \{0\}$, $\mathrm{in}_w(I)$ can also be defined as the additive group generated by $\{\mathrm{init}_w(f) : f \in I\} \subset R$.

3.2.1 Weights and Monomial Orders

The passage from an ideal I to $\mathrm{in}_w(I)$ can be seen as a "continuous" degenerative process. Before explaining this, we will see that initial ideals with respect to weights are more general than initial ideals with respect to monomial orders. Precisely, we will show that, fixed a monomial order $<$ on R and an ideal $I \subset R$, we can always find a suitable $w \in (\mathbb{N} \setminus \{0\})^n$ such that $\mathrm{in}_w(I) = \mathrm{in}_<(I)$.

Example 3.8 Let us find a weight vector $w = (a, b, c) \in (\mathbb{N} \setminus \{0\})^3$ that picks the largest monomial in every subset of monomials of degree $\leq d$ in $K[X, Y, Z]$ (with respect to the standard grading) for the lexicographic order determined by $X > Y > Z$. We choose $c = 1$. Since $Y > Z^d$ and $w(Z^d) = d$, we choose $b = d + 1$. Since $X > Y^d$ and $w(Y^d) = d(d + 1)$, we choose $a = d(d + 1) + 1$. It is not hard to check that $w = (d(d + 1) + 1, d + 1, 1)$ indeed solves our problem.

With this choice, if $I \subset K[X, Y, Z]$ is an ideal admitting a Gröbner basis of polynomials of degree at most d, then $\text{in}(I) = \text{in}_w(I)$.

The last paragraph of the above example will be formalized in the next lemma. Given $w \in \mathbb{N}^n$ and a monomial order $<$, we define another monomial order on $R = K[X_1, \ldots, X_n]$ by

$$\mu <_w \nu \iff \begin{array}{l} w(\mu) < w(\nu) \quad \text{or} \\ w(\mu) = w(\nu) \text{ and } \mu < \nu \end{array}.$$

Lemma 3.2 *For an ideal* $I \subset R = K[X_1, \ldots, X_n]$, *if* $\text{in}_w(I) \subset \text{in}_<(I)$ *or* $\text{in}_w(I) \supset \text{in}_<(I)$, *then* $\text{in}_w(I) = \text{in}_<(I)$.

Proof By applying $\text{in}_<(-)$ on both sides of the inclusion $\text{in}_w(I) \supset \text{in}_<(I)$ we get

$$\text{in}_{<_w}(I) = \text{in}_<(\text{in}_w(I)) \supset \text{in}_<(\text{in}_<(I)) = \text{in}_<(I).$$

So the equality $\text{in}_<(\text{in}_w(I)) = \text{in}_<(\text{in}_<(I))$ must hold by Corollary 3.1, and since $\text{in}_w(I) \supset \text{in}_<(I)$ we must have $\text{in}_w(I) = \text{in}_<(I)$ by Corollary 3.2.

The case in which $\text{in}_w(I) \subset \text{in}_<(I)$ is analogous. □

The next lemma explains why the initial monomial of a nonzero polynomial $f \in R$ must correspond to a vertex of the Newton polytope $\text{NP}(f) \subset \mathbb{R}^n$.

Lemma 3.3 *Let* $P \subset \mathbb{R}^n$ *be the convex hull of some vectors* $\mathbf{u}^1, \ldots, \mathbf{u}^m \in \mathbb{N}^n$. *For a given monomial order* $<$ *on* $R = K[X_1, \ldots, X_n]$ *we have* $\mathbf{X}^{\mathbf{u}} \leq \max\{\mathbf{X}^{\mathbf{u}^1}, \ldots, \mathbf{X}^{\mathbf{u}^m}\}$ *for any* $\mathbf{u} \in P \cap \mathbb{N}^n$.

Proof If $\mathbf{u} \in P \cap \mathbb{N}^n$, then $\mathbf{u} = \sum_{i=1}^m \lambda_i \mathbf{u}^i$ with $\lambda_i \in \mathbb{Q}_{\geq 0}$ and $\sum_{i=1}^m \lambda_i = 1$ (for the reason why the λ_i's can be taken rational see, for example, [3, Proposition 1.70]). If $\lambda_i = a_i/b_i$ with $a_i \in \mathbb{N}, b_i \in \mathbb{N} \setminus \{0\}$ then, setting $b := b_1 \cdots b_m$ and $a_i' := a_i(b/b_i)$, we have

$$b\mathbf{u} = \sum_{i=1}^m a_i' \mathbf{u}^i.$$

If, by contradiction, $\mathbf{X}^{\mathbf{u}} > \mathbf{X}^{\mathbf{u}^i}$ for all $i = 1, \ldots, m$, since $b = \sum_{i=1}^m a_i'$ we would have

$$(\mathbf{X}^{\mathbf{u}})^b > (\mathbf{X}^{\mathbf{u}^1})^{a_1'} \cdots (\mathbf{X}^{\mathbf{u}^m})^{a_m'}$$

but this contradicts the fact that these two monomials are the same. □

Proposition 3.4 *Given a monomial order $<$ on $R = K[X_1, \ldots, X_n]$ and $\mu_i, v_i \in$ Mon(R) such that $\mu_i > v_i$ for $i = 1, \ldots, k$, there exists $w \in (\mathbb{N} \setminus \{0\})^n$ such that $w(\mu_i) > w(v_i) \ \forall \ i = 1, \ldots, k$. Consequently, given an ideal $I \subset R$ there exists $w \in (\mathbb{N} \setminus \{0\})^n$ such that $\text{in}_<(I) = \text{in}_w(I)$.*

Proof Notice that, for all $i = 1, \ldots, k$,

- $\mu_i > v_i \iff \prod_j \mu_j > v_i \prod_{j \neq i} \mu_j$;
- $w(\mu_i) > w(v_i) \iff w(\prod_j \mu_j) > w(v_i \prod_{j \neq i} \mu_j)$,

so, without loss of generality, we can assume that μ_i is the same monomial μ for all $i = 1, \ldots, k$.

If $\mu = \mathbf{X}^{\mathbf{u}}$ and $v_i = \mathbf{X}^{\mathbf{v}^i}$, consider the cone $C = \mathbf{u} + (\mathbb{R}_{\geq 0})^n \subset \mathbb{R}^n$ and the convex hull $P \subset \mathbb{R}^n$ of $\{\mathbf{u}, \mathbf{v}^1, \ldots, \mathbf{v}^k\}$. We claim that $C \cap P = \{\mathbf{u}\}$. Suppose that $\mathbf{v} \in C \cap P$. It is harmless to assume that $\mathbf{v} \in \mathbb{Q}^n$, so that there is $N \in \mathbb{N}$ big enough such that $N\mathbf{v} \in \mathbb{N}^n$. Let $v = \mathbf{X}^{N\mathbf{v}}$. Since $\mathbf{v} \in C$, v is divided by $\mu^N = \mathbf{X}^{N\mathbf{u}}$, so $v \geq \mu^N$. On the other hand, $\mathbf{v} \in P \Rightarrow N\mathbf{v} \in NP = \{N\mathbf{z} : \mathbf{z} \in P\}$, so by Lemma 3.3 we have $v \leq \max\{\mathbf{X}^{N\mathbf{u}}, \mathbf{X}^{N\mathbf{v}^i} : i = 1, \ldots, k\} = \mathbf{X}^{N\mathbf{u}}$, hence $v = \mu^N$. That is: $\mathbf{v} = \mathbf{u}$.

Therefore there is a hyperplane passing through $\mathbf{u}$ separating C and P (for this see, for example, [3, Theorem 1.32]), that is there is $w \in (\mathbb{R}^n)^*$ such that

$$w(\mathbf{v}) > w(\mathbf{u}) > w(\mathbf{v^i})$$

for all $\mathbf{v} \in C \setminus \{\mathbf{u}\}$ and $i = 1, \ldots, k$. Thinking of w as a real vector $w = (w_1, \ldots, w_n)$, it is harmless to assume $w_j \in \mathbb{Q}$ for all j; furthermore the first inequalities yield $w_j > 0$ for all $j = 1, \ldots, n$. Hence, after taking a suitable multiple, we can assume that $w \in (\mathbb{N} \setminus \{0\})^n$ is our desired weight vector.

For the last part of the statement, let $f_1, \ldots, f_m$ be a Gröbner basis of I. By the first part, there is $w \in (\mathbb{N} \setminus \{0\})^n$ such that $w(\mu) > w(v)$ where $\mu = \text{in}(f_i)$ and $v \in \text{supp}(f_i) \setminus \{\mu\}$ for all $i = 1, \ldots, m$. So $\text{in}_<(I) \subset \text{in}_w(I)$, hence $\text{in}_<(I) = \text{in}_w(I)$ by Lemma 3.2. □

3.2.2 Degeneration to the Initial Ideal

Let us extend $R = K[X_1, \ldots, X_n]$ to $P := R[t]$ by introducing a *homogenizing variable* t. The *w-homogenization* of $f = \sum_{\mu \in \text{supp}(f)} a_\mu \mu \in R$ is

$$\text{hom}_w(f) := \sum_{\mu \in \text{supp}(f)} a_\mu \mu t^{w(f) - w(\mu)} \in P.$$

Example 3.9 Let $f = X^2 - XY + Z^2 \in K[X, Y, Z]$. We have:

- $\hom_w(f) = X^2 - XY + Z^2 t^2$ if $w = (2, 2, 1)$.
- $\hom_w(f) = X^2 - XY t^2 + Z^2 t^6$ if $w = (4, 2, 1)$.

Given an ideal $I \subset R$, $\hom_w(I) \subset P$ denotes the ideal generated by $\hom_w(f)$ with $f \in I$. Once again, $\hom_w(I)$ can be defined as the additive group generated by $\hom_w(f)$ with $f \in I$. For its study, we extend the weight vector w to w' on P by $w_i' = w_i$ for all $i \leq n$ and $w_{n+1}' = 1$, so that $\hom_w(I)$ is a w'-homogeneous ideal of P, where the grading is $\deg(X_i) = w_i$ and $\deg(t) = 1$. Since $P/\hom_w(I)$ is a w'-graded P-module, it is also a graded $K[t]$-module with respect to the standard grading on $K[t]$. So $t - a$ is not a zero-divisor on $P/\hom_w(I)$ for any $a \in K \setminus \{0\}$. We want to show that t is not a zero-divisor on $P/\hom_w(I)$ as well, and in order to do so it is useful to introduce the *dehomogenization map*:

$$\pi : P \longrightarrow R$$

$$F(X_1, \ldots, X_n, t) \mapsto F(X_1, \ldots, X_n, 1).$$

Remark 3.5 We want to highlight the following easy facts:

1. $\pi(\hom_w(f)) = f \ \forall \ f \in R$. So, $\pi(\hom_w(I)) = I$.
2. If $F \in P \setminus tP$ is w'-homogeneous, then $\hom_w(\pi(F)) = F$; moreover, if $r \in \mathbb{N}$ and $G = t^r F$, $\hom_w(\pi(G)) t^r = G$.

Summarizing, for any $F \in P$ we have $F \in \hom_w(I) \iff \pi(F) \in I$.

Proposition 3.5 *Given an ideal I of R, the element $t - a \in K[t]$ is not a zero divisor on $P/\hom_w(I)$ for every $a \in K$. Furthermore:*

- $P/\big(\hom_w(I) + (t)\big) \cong R/\mathrm{in}_w(I)$.
- $P/\big(\hom_w(I) + (t - a)\big) \cong R/I$ *for all $a \in K \setminus \{0\}$.*

Proof The first assertion needs to be proved just for $a = 0$. Let $F \in P$ such that $tF \in \hom_w(I)$. Using Remark 3.5 we infer that $\pi(tF) \in I$ and so, since $\pi(F) = \pi(tF)$, $F \in \hom_w(I)$.

For $P/(\hom_w(I) + (t)) \cong R/\mathrm{in}_w(I)$ it is enough to check that $\hom_w(I) + (t) = \mathrm{in}_w(I) + (t)$. This is easily seen since for every $f \in R$ the difference $\hom_w(f) - \mathrm{init}_w(f)$ is divisible by t.

Finally, to prove that $P/(\hom_w(I) + (t - a)) \cong R/I$ for every $a \in K \setminus \{0\}$, we consider the graded isomorphism $\psi : R \to R$ induced by $\psi(X_i) = a^{-w_i} X_i$. Of course $\psi(\mu) = a^{-w(\mu)} \mu \ \forall \ \mu \in \mathrm{Mon}(R)$ and $\hom_w(f) - a^{w(f)} \psi(f)$ is divisible by $t - a$ for all $f \in R$. So $\hom_w(I) + (t - a) = \psi(I) + (t - a)$, which implies the desired isomorphism. $\qquad\square$

Remark 3.6 Since a module over a principal ideal domain is flat if and only if it has no torsion (for example, this follows by [14, Theorem 7.7]), the proposition above says that $P/\hom_w(I)$ is a flat $K[t]$-module, and that it defines a flat family over $K[t]$ with generic fibre R/I and special fibre $R/\mathrm{in}_w(I)$.

3.2.2.1 Semicontinuity

Lemma 3.4 *Let A be a ring, M, N A-modules and $a \in \mathrm{ann}(N) \subset A$ a nonzerodivisor on M as well as on A. Then, for all $i \geq 0$,*

$$\mathrm{Ext}_A^i(M, N) \cong \mathrm{Ext}_{A/aA}^i(M/aM, N).$$

Proof Let $F_\bullet$ be a free resolution of M. The Ext modules on the left hand side are the cohomology modules of $\mathrm{Hom}_A(F_\bullet, N)$, which is a complex of A-modules isomorphic to $\mathrm{Hom}_{A/aA}(F_\bullet/aF_\bullet, N)$ because a annihilates N. However $F_\bullet/aF_\bullet$ is a free resolution of the A/aA-module M/aM since a is a nonzerodivisor on M as well as on A, so the cohomology modules of the latter complex are the Ext modules on the right hand side.   □

Let us remind that $R = K[X_1, \ldots, X_n]$ is graded by $\deg(X_i) = g_i$ where $g = (g_1, \ldots, g_n)$ is a vector of positive integers. In this way $\mathfrak{m} = (X_1, \ldots, X_n)$ is the unique homogeneous maximal ideal of R. If $I \subset R$ is a homogeneous ideal, then $\hom_w(I) \subset P$ is homogeneous with respect to the *bi-graded* structure on $P = R[t]$ given by $\deg(X_i) = (g_i, w_i)$ and $\deg(t) = (0, 1)$. So $S := P/\hom_w(I)$ and $\mathrm{Ext}_P^i(S, P)$ are finitely generated bi-graded P-modules.

Remark 3.7 Given a bi-graded P-module M, for all $j \in \mathbb{Z}$ we can consider the $K[t]$-module $M_{(j,*)} := \bigoplus_{k \in \mathbb{Z}} M_{(j,k)}$: clearly $M_{(j,*)}$ is graded with respect to the standard grading of $K[t]$, the piece of degree k being $(M_{(j,*)})_k = M_{(j,k)}$ for all $k \in \mathbb{Z}$.

Furthermore, if M is finitely generated over P, then $M_{(j,*)}$ is finitely generated over $K[t]$ for all $j \in \mathbb{Z}$. In fact, we consider the P-submodule $\bigoplus_{i \geq j} M_{(i,*)} \subset M$. Such a module is finitely generated over P, and being bi-graded the finite generating set can be chosen to be consisting of bi-homogeneous elements. It is trivial to check that, among such finitely many generators, the ones of bi-degree $(j, *)$ generate $M_{(j,*)}$ as a $K[t]$-module.

Remark 3.8 If N is a finitely generated $K[t]$-module, by the structure theorem for finitely generated modules over a principal ideal domain (like $K[t]$), we have

$$N \cong K[t]^a \oplus \left(\bigoplus_{i=1}^r K[t]/(f_i) \right)$$

for $a, r \in \mathbb{N}$ and polynomials $f_1, \ldots f_r \in K[t]$. If N is also graded, then each f_i must be homogeneous, and so $f_i = t^{k_i}$ for a positive integer k_i.

Let I be a homogeneous ideal of R and $S = P/\hom_w(I)$. From the above discussion and remarks, for all $i, j \in \mathbb{Z}$ we have

$$\mathrm{Ext}_P^i(S, P)_{(j,*)} \cong K[t]^{a_{i,j}} \oplus \left(\bigoplus_{k \in \mathbb{N} \setminus \{0\}} (K[t]/(t^k))^{b_{i,j,k}} \right)$$

for some natural numbers $a_{i,j}$ and $b_{i,j,k}$. Let $b_{i,j} = \sum_{k\in\mathbb{N}\setminus\{0\}} b_{i,j,k}$. In the following, $H^i_\mathfrak{m}(M)$ denotes the ith *local cohomology module* of an R-module M with support in the ideal $\mathfrak{m} = (X_1, \dots, X_n)$. We also recall that, if M is graded and finitely generated, the *depth* of M is:

$$\mathrm{depth}M = \inf\{i \in \mathbb{N} : H^i_\mathfrak{m}(M) \neq 0\}$$

and its *Castelnuovo-Mumford regularity* is:

$$\mathrm{reg}M = \sup\{i + j \in \mathbb{N} : H^i_\mathfrak{m}(M)_j \neq 0\}.$$

Theorem 3.1 *With the above notation, for any $i, j \in \mathbb{Z}$ we have:*

- $\dim_K(\mathrm{Ext}^i_R(R/I, R)_j) = a_{i,j}$.
- $\dim_K(\mathrm{Ext}^i_R(R/\mathrm{in}_w(I), R)_j) = a_{i,j} + b_{i,j} + b_{i+1,j}$.

Therefore, $\dim_K(\mathrm{Ext}^i_R(R/I, R)_j) \leq \dim_K(\mathrm{Ext}^i_R(R/\mathrm{in}_w(I), R)_j)$ *and, equivalently,* $\dim_K(H^i_\mathfrak{m}(R/I)_j) \leq \dim_K(H^i_\mathfrak{m}(R/\mathrm{in}_w(I))_j)$. *In particular* $\mathrm{depth}R/\mathrm{in}_w(I) \leq \mathrm{depth}R/I$ *and* $\mathrm{reg}R/\mathrm{in}_w(I) \geq \mathrm{reg}R/I$.

Proof Letting x be t or $t - 1$, we have the short exact sequence

$$0 \to P \xrightarrow{\cdot x} P \to P/xP \to 0.$$

The long exact sequence of $\mathrm{Ext}_P(S, -)$ associated to it gives us the following short exact sequences of P-modules for all $i \in \mathbb{Z}$:

$$0 \to \mathrm{Coker}\alpha_{i,x} \to \mathrm{Ext}^i_P(S, P/xP) \to \mathrm{Ker}\alpha_{i+1,x} \to 0,$$

where $\alpha_{k,x}$ is the multiplication by x on $\mathrm{Ext}^k_P(S, P)$. We can restrict the above to the degree $(j, *)$ for any $j \in \mathbb{Z}$ getting the short exact sequences of $K[t]$-modules:

$$0 \to (\mathrm{Coker}\alpha_{i,x})_{(j,*)} \to (\mathrm{Ext}^i_P(S, P/xP))_{(j,*)} \to (\mathrm{Ker}\alpha_{i+1,x})_{(j,*)} \to 0.$$

Notice that we have:

- $(\mathrm{Coker}\alpha_{i,t})_{(j,*)} \cong K^{a_{i,j}+b_{i,j}}$ and $(\mathrm{Ker}\alpha_{i+1,t})_{(j,*)} \cong K^{b_{i+1,j}}$.
- $(\mathrm{Coker}\alpha_{i,t-1})_{(j,*)} \cong K^{a_{i,j}}$ and $(\mathrm{Ker}\alpha_{i+1,t-1})_{(j,*)} = 0$.

Therefore, for all $i, j \in \mathbb{Z}$, we get:

- $(\mathrm{Ext}^i_P(S, P/tP))_{(j,*)} \cong K^{a_{i,j}+b_{i,j}+b_{i+1,j}}$.
- $(\mathrm{Ext}^i_P(S, P/(t - 1)P))_{(j,*)} \cong K^{a_{i,j}}$.

By Lemma 3.4 and Proposition 3.5 we get the isomorphisms of K-vector spaces:

- $(\mathrm{Ext}^i_P(S, P/tP))_{(j,*)} \cong (\mathrm{Ext}^i_{P/tP}(S/tS, P/tP))_{(j,*)} \cong (\mathrm{Ext}^i_R(R/\mathrm{in}_w(I), R))_j$.
- $(\mathrm{Ext}^i_P(S, P/(t - 1)P))_{(j,*)} \cong (\mathrm{Ext}^i_{P/(t-1)P}(S/(t - 1)S, P/(t - 1)P))_{(j,*)} \cong (\mathrm{Ext}^i_R(R/I, R))_j$.

The thesis follows from this. For the local cohomology statement just observe that by Grothendieck graded duality (for example see [4, Theorem 3.19]) $H^i_{\mathfrak{m}}(R/J)_j$ is dual as a K-vector space to $\text{Ext}^{n-i}_R(R/J, R)_{-|g|-j}$ for any homogeneous ideal $J \subset R$ and $i, j \in \mathbb{Z}$ (here $|g| = g_1 + \ldots + g_n$). $\qquad\qquad\qquad\square$

3.3 Fibre-Full Modules

The main purpose of this section is to prove Theorem 3.2, which naively says that some homological properties are preserved along a flat family provided the special fibre is good. The argument given here is inspired by the work by Kollár and Kovács [9]. We point out that, when A is Gorenstein, Theorem 3.2 could also be proved via the techniques used by Ma and Pham in [12, Proposition 3.3], exploiting Grothendieck local duality.

Throughout this section, A is a Noetherian flat $K[t]$-algebra and M a finitely generated A-module which is flat over $K[t]$. Also, both A and M are graded $K[t]$-modules with respect to the standard grading of $K[t]$, and we assume that $1 \in A_0$ and $A_i = 0$ for all $i < 0$ (think of A and M as if they were, with the notation of Sect. 3.2.2.1, respectively P and S).

Remark 3.9 Given a base ring B and a B-module L, we recall that L is flat over B if and only if the natural map $I \otimes_B L \to L$ sending $x \otimes m$ to xm is injective for any ideal $I \subset B$ (cf. [14, Theorem 7.7]). In particular:

- With the notation just introduced, the fact that the graded modules A and M are flat over the principal ideal domain $K[t]$ is equivalent to the fact that t is a nonzero divisor both on A and on M.
- For a positive integer m, a $K[t]/(t^m)$-module L is flat over $K[t]/(t^m)$ if and only if $0 :_L t^k = t^{m-k}L$ for all $k = 0, \ldots, m$.

Lemma 3.5 *The following are equivalent:*

1. $\text{Ext}^i_A(M, A)$ *is flat over* $K[t]$ *for all* $i \in \mathbb{N}$.
2. $\text{Ext}^i_{A/t^m A}(M/t^m M, A/t^m A)$ *is flat over* $K[t]/(t^m)$ $\forall\, i \in \mathbb{N}, m \in \mathbb{N} \setminus \{0\}$.

Proof

$(1) \implies (2)$ Since A is flat over $K[t]$, there is a short exact sequence $0 \to A \xrightarrow{\cdot t^m} A \to A/t^m A \to 0$. Consider the induced long exact sequence of $\text{Ext}^\bullet_A(M, -)$:

$$\cdots \to \text{Ext}^i_A(M, A) \xrightarrow{\cdot t^m} \text{Ext}^i_A(M, A) \to \text{Ext}^i_A(M, A/t^m A)$$

$$\to \text{Ext}^{i+1}_A(M, A) \xrightarrow{\cdot t^m} \text{Ext}^{i+1}_A(M, A) \to \cdots$$

Since $\mathrm{Ext}_A^k(M, A)$ does not have t-torsion for all $k \in \mathbb{N}$, for all $i \in \mathbb{N}$ we have a short exact sequence

$$0 \to \mathrm{Ext}_A^i(M, A) \xrightarrow{\cdot t^m} \mathrm{Ext}_A^i(M, A) \to \mathrm{Ext}_A^i(M, A/t^m A) \to 0,$$

from which $\mathrm{Ext}_A^i(M, A/t^m A) \cong \dfrac{\mathrm{Ext}_A^i(M, A)}{t^m \mathrm{Ext}_A^i(M, A)}$. It is straightforward to check that the latter is flat over $K[t]/(t^m)$ because of (1). Finally, Lemma 3.4 implies that

$$\mathrm{Ext}_A^i(M, A/t^m A) \cong \mathrm{Ext}_{A/t^m A}^i(M/t^m M, A/t^m A).$$

(2) $\implies$ (1) By contradiction, suppose $\mathrm{Ext}_A^i(M, A)$ is not flat over $K[t]$. Then there exists a nontrivial class $[\phi] \in \mathrm{Ext}_A^i(M, A)$ such that $t[\phi] = 0$.

Let us take an A-free resolution $F_\bullet$ of M, and let $(G^\bullet, \partial^\bullet)$ be the complex $\mathrm{Hom}_A(F_\bullet, A)$, so that $\mathrm{Ext}_A^i(M, A)$ is the ith cohomology module of $G^\bullet$. Then $\phi \in \mathrm{Ker}(\partial^i) \setminus \mathrm{Im}(\partial^{i-1})$ and $t\phi \in \mathrm{Im}(\partial^{i-1})$.

Since M and A are flat over $K[t]$, $F_\bullet/t^m F_\bullet$ is a $A/t^m A$-free resolution of $M/t^m M$. Let $(\overline{G}^\bullet, \overline{\partial}^\bullet)$ denote the complex $\mathrm{Hom}_{A/t^m A}(F_\bullet/t^m F_\bullet, A/t^m A)$, so that

$$\mathrm{Ext}_{A/t^m A}^i(M/t^m M, A/t^m A) \cong H^i(\overline{G}^\bullet).$$

Moreover let $\pi^\bullet$ be the obvious map of complexes from $G^\bullet$ to $\overline{G}^\bullet$. Of course $\pi^i(\phi) \in \mathrm{Ker}(\overline{\partial^i})$ and $t\pi^i(\phi) \in \mathrm{Im}(\overline{\partial^{i-1}})$. It is enough to find a positive integer m such that $\pi^i(\phi)$ does not belong to $\mathrm{Im}(\overline{\partial^{i-1}}) + t^{m-1}\mathrm{Ker}(\overline{\partial^i})$. Indeed, in this case $x = [\pi^i(\phi)]$ would be an element of $\mathrm{Ext}_{A/t^m A}^i(M/t^m M, A/t^m A)$ killed by t but not in $t^{m-1}\mathrm{Ext}_{A/t^m A}^i(M/t^m M, A/t^m A)$ contradicting (2).

If $\pi^i(\phi) \in \mathrm{Im}(\overline{\partial^{i-1}}) + t^{m-1}\mathrm{Ker}(\overline{\partial^i})$, then $\phi \in \mathrm{Im}(\partial^{i-1}) + t^{m-1}\mathrm{Ker}(\partial^i) + t^m G^i$. Hence $\phi = \alpha + t^{m-1}\beta + t^m \gamma$ with $\alpha \in \mathrm{Im}(\partial^{i-1})$, $\beta \in \mathrm{Ker}(\partial^i)$ and $\gamma \in G^i$. Moreover, since $\phi \in \mathrm{Ker}(\partial^i)$ and t is A-regular, we have $\gamma \in \mathrm{Ker}(\partial^i)$. Therefore we deduce that

$$\phi \in \mathrm{Im}(\partial^{i-1}) + t^{m-1}\mathrm{Ker}(\partial^i).$$

If this happened for all $m \gg 0$, by Krull's intersection theorem (cf. [14, Theorem 8.9]) there exists $a \in A$ such that $1 + ta$ kills $[\phi] \in \mathrm{Ext}_A^i(M, A)$; since $t[\phi] = 0$, then $[\phi] = 0$, which is a contradiction. Therefore $\pi^i(\phi)$ is not in $\mathrm{Im}(\overline{\partial^{i-1}}) + t^{m-1}\mathrm{Ker}(\overline{\partial^i})$ for $m \gg 0$. $\qquad\square$

Lemma 3.5 ensures that to show that $\mathrm{Ext}_A^i(M, A)$ is flat over $K[t]$ for all $i \in \mathbb{N}$ it is enough to show that the $K[t]/(t^m)$-module $\mathrm{Ext}_{A/t^m A}^i(M/t^m M, A/t^m A)$ is flat for all $i \in \mathbb{N}$ and $m \in \mathbb{N} \setminus \{0\}$. We are going to show this under a further assumption on M:

Definition 3.8 Let $K[t]$, A and M as above. We say that M is *fibre-full* if, for all $m \geq 1$, the natural surjective map $M/t^m M \to M/tM$ induces injective maps $\mathrm{Ext}_A^i(M/tM, A) \to \mathrm{Ext}_A^i(M/t^m M, A)$ for all $i \in \mathbb{N}$.

To our goal, it is helpful to introduce the following notation for all $m \geq 1$:

- $A_m = A/t^m A$.
- $M_m = M/t^m M$.
- $\iota_j : t^{j+1} M_m \to t^j M_m$ the natural inclusion for all $j = 0, \ldots, m-1$.
- $\mu_j : t^j M_m \to t^{m-1} M_m$ the multiplication by t^{m-1-j} for all $j = 0, \ldots, m-1$.
- $E_m^i(-)$ the contravariant functor $\mathrm{Ext}_{A_m}^i(-, A_m)$ for all $i \in \mathbb{N}$.

Remark 3.10 Whenever $k = 0, \ldots, m$, a result of Rees (cf. [4, Lemma 3.1.16]) implies that $E_m^i(M_k) \cong \mathrm{Ext}_A^{i+1}(M_k, A)$. Hence we deduce that

$$E_k^i(M_k) \cong E_m^i(M_k) \ \forall \ k = 0, \ldots, m.$$

Remark 3.11 Since t is a non-zero-divisor on M, the multiplication by t^{m-j} sending $x + t^j M$ to $t^{m-j} x + t^m M$ for all $x \in M$ gives an isomorphism:

$$M_j \cong t^{m-j} M_m \ \forall j = 0, \ldots, m.$$

Remark 3.12 The short exact sequences $0 \to t^{j+1} M_m \xrightarrow{\iota_j} t^j M_m \xrightarrow{\mu_j} t^{m-1} M_m \to 0$, if M is fibre-full, yield the following short exact sequences for all $i \in \mathbb{N}$:

$$0 \to E_m^i(t^{m-1} M_m) \xrightarrow{E_m^i(\mu_j)} E_m^i(t^j M_m) \xrightarrow{E_m^i(\iota_j)} E_m^i(t^{j+1} M_m) \to 0.$$

Indeed, up to the identifications of Remark 3.11, μ_j corresponds to the natural projection $M_{m-j} \to M_1$, therefore the map $E_m^i(\mu_j)$ is injective for all $i \in \mathbb{N}$ by definition of fibre-full module.

Theorem 3.2 *With the above notation, if M is a fibre-full A-module, then $\mathrm{Ext}_A^i(M, A)$ is flat over $K[t]$ for all $i \in \mathbb{N}$.*

Proof By Lemma 3.5, it is enough to show that $E_m^i(M_m)$ is flat over $K[t]/(t^m)$ for all $m \in \mathbb{N} \setminus \{0\}$. This is clear for $m = 1$ (because $K[t]/(t)$ is a field), so we will proceed by induction: thus let us fix $m \geq 2$ and assume that $E_{m-1}^i(M_{m-1})$ is flat over $K[t]/(t^{m-1})$. The local flatness criterion (cf. [14, Theorem 22.3]) tells us that is enough to show the following two properties:

1. $E_m^i(M_m)/t^{m-1} E_m^i(M_m)$ is flat over $K[t]/(t^{m-1})$.
2. The map $(t^{m-1})/(t^m) \otimes_{K[t]/(t^m)} E_m^i(M_m) \to E_m^i(M_m)$ sending $\overline{t^{m-1}} \otimes \phi$ to $t^{m-1}\phi$ is injective: in other words, $0 :_{E_m^i(M_m)} t^{m-1} = \mathrm{Ker}(E_m^i(\iota^0)) = t E_m^i(M_m)$.

By Remark 3.12 $E_m^i(\iota_k)$ is surjective for all k. Since $E_m^i(-)$ is a functor, and the composition of surjective maps is surjective, we infer that $E_m^i(\iota^j)$ is surjective where

$$\iota^j := \iota_j \circ \ldots \iota_{m-2} : t^{m-1} M_m \to t^j M_m.$$

Since $\iota^j \circ \mu_j$ is the multiplication by t^{m-1-j} on $t^j M_m$, and $E_m^i(-)$ is a A_m-linear functor, we therefore have

$$\mathrm{Im}(E_m^i(\mu_j)) = \mathrm{Im}(E_m^i(\mu_j) \circ E_m^i(\iota^j)) = \mathrm{Im}(E_m^i(\iota^j \circ \mu_j)) = t^{m-1-j} E_m^i(t^j M_m).$$

Therefore $\mathrm{Ker}(E_m^i(\iota_j)) = t^{m-1-j} E_m^i(t^j M_m)$. Hence

$$E_m^i(t^{j+1} M_m) \cong \frac{E_m^i(t^j M_m)}{t^{m-1-j} E_m^i(t^j M_m)}.$$

Plugging in $j = 0$, we get that

$$\frac{E_m^i(M_m)}{t^{m-1} E_m^i(M_m)} \cong E_m^i(t M_m) \cong E_m^i(M_{m-1}) \cong E_{m-1}^i(M_{m-1})$$

(where the last two isomorphisms follow by Remarks 3.11 and 3.10) is flat over $K[t]/(t^{m-1})$ by induction, and this shows 1.

Concerning 2, *we first claim that the kernel of* $E_m^i(\iota^0)$: $E_m^i(M_m) \to E_m^i(t^{m-1} M_m)$ *is equal to* $t E_m^i(M_m)$: indeed, we show by reverse induction on $j = 0, \ldots, m-2$ that $\mathrm{Ker}(E_m^i(\iota^j)) = t E_m^i(t^j M_m)$, being this already proved for $j = t - 2$ (because $E_m^i(\iota^{m-2}) = E_m^i(\iota_{m-2})$). Obviously $t E_m^i(t^j M_m) \subset \mathrm{Ker}(E_m^i(\iota^j))$; on the other hand, if $u \in \mathrm{Ker}(E_m^i(\iota^j))$, by induction $E_m^i(\iota_j)(u) = t v$ for some $v \in E_m^i(t^{j+1} M_m)$. Being $E_m^i(\iota_j)$ surjective, there is $u' \in E_m^i(t^j M_m)$ such that $E_m^i(\iota_j)(u') = v$. But then $u - t u' \in \mathrm{Ker}(E_m^i(\iota_j)) = t^{m-1-j} E_m^i(t^j M_m)$, and so $u \in t E_m^i(t^j M_m)$. This proves the claim.

Now, since $E_m^i(\mu_0) \circ E_m^i(\iota^0)$ is the multiplication by t^{m-1} on $E_m^i(M_m)$ and $E_m^i(\mu_0)$ is an injective map, we get

$$0 :_{E_m^i(M_m)} t^{m-1} = \mathrm{Ker}(E_m^i(\iota^0)) = t E_m^i(M_m),$$

which concludes the proof of 2. $\qquad\square$

As a consequence of Theorem 3.2, exploiting Theorem 3.1, we have:

Corollary 3.3 *Let* $I \subset R = K[X_1, \ldots, X_n]$ *be an ideal such that* $S = P/\hom_w(I)$ *is a fibre-full* P-module $(P = R[t])$. *Then* $\mathrm{Ext}_P^i(S, P)$ *is a flat* $K[t]$-*module. So, if furthermore* I *is homogeneous:*

$$\dim_K(H_{\mathrm{m}}^i(R/I)_j) = \dim_K(H_{\mathrm{m}}^i(R/\mathrm{in}_w(I))_j) \ \forall \, i, j \in \mathbb{Z}.$$

In particular $\mathrm{depth} R/\mathrm{in}_w(I) = \mathrm{depth} R/I$ *and* $\mathrm{reg} R/\mathrm{in}_w(I) = \mathrm{reg} R/I$.

3.4   Squarefree Monomial Ideals and Fibre-Full Modules

Because of Corollary 3.3, our next goal is to show that, for an ideal $I \subset R$ such that $\mathrm{in}_w(I)$ is a squarefree monomial ideal, then $S = P/\mathrm{hom}_w(I)$ is a fibre-full P-module. To do so, we need to recall some notion.

Let $J \subset R$ be a monomial ideal minimally generated by monomials $\mu_1, \ldots, \mu_r$. For every subset $\sigma \subset \{1, \ldots, r\}$ we define the monomial $\mu(J, \sigma) := \mathrm{lcm}(\mu_i | i \in \sigma) \in R$. If v is the qth element of σ we set $\mathrm{sign}(v, \sigma) := (-1)^{q-1} \in K$. Let us consider the graded complex of free R-modules $F_\bullet(J) = (F_i, \partial_i)_{i=0,\ldots,r}$ with

$$F_i := \bigoplus_{\substack{\sigma \subset \{1,\ldots,r\} \\ |\sigma|=i}} R(-d_\sigma), \quad d_\sigma = \deg \mu(J, \sigma),$$

and differentials defined by

$$1_\sigma \mapsto \sum_{v \in \sigma} \mathrm{sign}(v, \sigma) \frac{\mu(J, \sigma)}{\mu(J, \sigma \setminus \{v\})} \cdot 1_{\sigma \setminus \{v\}}.$$

It is well known that $F_\bullet(J)$ is a graded free R-resolution of R/J (cf. [1, Remark 2.3]), called the *Taylor resolution*.

For any positive integer k we introduce the monomial ideal

$$J^{[k]} := (\mu_1^k, \ldots, \mu_r^k).$$

Notice that $\mu_1^k, \ldots, \mu_r^k$ is the minimal system of monomial generators of $J^{[k]}$, so $\mu(J^{[k]}, \sigma) = \mu(J, \sigma)^k$ for any $\sigma \subset \{1, \ldots, r\}$.

Theorem 3.3 *If $J \subset R$ is a squarefree monomial ideal, for all $i \in \mathbb{Z}$ and $k \in \mathbb{N} \setminus \{0\}$ the map $\mathrm{Ext}_R^i(R/J^{[k]}, R) \to \mathrm{Ext}_R^i(R/J^{[k+1]}, R)$ induced by the projection $R/J^{[k+1]} \to R/J^{[k]}$ is injective.*

Corollary 3.4 *Let $I \subset R = K[X_1, \ldots, X_n]$ be an ideal such that $\mathrm{in}_w(I)$ is a squarefree monomial ideal. Then $S = P/\mathrm{hom}_w(I)$ is a fibre-full P-module.*

Proof of Corollary 3.4 Notice that $\mathrm{hom}_w(I) + tP = \mathrm{in}_w(I) + tP$ is a squarefree monomial ideal of P. So, by Theorem 3.3, the maps $\mathrm{Ext}_P^i(S/tS, P) \to \mathrm{Ext}_P^i(P/(\mathrm{hom}_w(I)+tP)^{[m]}, P)$ are injective for all $m \in \mathbb{N} \setminus \{0\}$. Since $(\mathrm{hom}_w(I) + tP)^{[m]} \subset (\mathrm{hom}_w(I) + tP)^m \subset \mathrm{hom}_w(I) + t^m P$, these maps factor through $\mathrm{Ext}_P^i(S/tS, P) \to \mathrm{Ext}_P^i(S/t^m S, P)$, hence the latter are injective as well.   □

Proof of Theorem 3.3 Let $u_1, \ldots, u_r$ be the minimal monomial generators of J. For all $k \in \mathbb{N} \setminus \{0\}$, $\sigma \subset \{1, \ldots, r\}$ set $\mu_\sigma[k] := \mu(J^{[k]}, \sigma)$ and $\mu_\sigma := \mu_\sigma[1]$. Of course μ_σ is a squarefree monomial and $\mu_\sigma[k] = \mu_\sigma^k$. The module $\mathrm{Ext}_R^i(R/J^{[k]}, R)$ is the ith cohomology of the complex $G^\bullet[k] = \mathrm{Hom}_R(F_\bullet[k], R)$ where $F_\bullet[k] = F_\bullet(J^{[k]}) = (F_i, \partial_i[k])_{i=0,\ldots,r}$ is the Taylor resolution of $R/J^{[k]}$. Let $F_i \xrightarrow{f_i} F_i$ be the map sending 1_σ to $\mu_\sigma \cdot 1_\sigma$. It is easy to check that the collection $F_\bullet[k+1] \xrightarrow{f_\bullet=(f_i)_i} F_\bullet[k]$ is a morphism of complexes lifting $R/J^{[k+1]} \to R/J^{[k]}$.

So the maps $\mathrm{Ext}^i_R(R/J^{[k]}, R) \to \mathrm{Ext}^i_R(R/J^{[k+1]}, R)$ we are interested in are the homomorphisms $H^i(G^\bullet[k]) \xrightarrow{\overline{g^i}} H^i(G^\bullet[k+1])$ induced by $g^\bullet = \mathrm{Hom}(f_\bullet, R)$: $G^\bullet[k] \to G^\bullet[k+1]$. Let us see how $\overline{g^i}$ acts: if $G^\bullet[k] = (G^i, \partial^i[k])$, then $G^i = \mathrm{Hom}_R(F_i, R)$ can be identified with F_i (ignoring the grading) and $\partial^i[k] : G^i \longrightarrow G^{i+1}$ sends

$$1_\sigma \mapsto \sum_{v \in \{1,\dots,r\}\setminus\sigma} \mathrm{sign}(v, \sigma \cup \{v\}) \left(\frac{\mu_{\sigma \cup \{v\}}}{\mu_\sigma} \right)^k \cdot 1_{\sigma \cup \{v\}}$$

for all $\sigma \subset \{1, \dots, r\}$ and $|\sigma| = i$. The map $g^i : G^i \to G^i$, up to the identification $F_i \cong G^i$, is then the map sending 1_σ to $\mu_\sigma \cdot 1_\sigma$.

We claim that $\overline{g^i}$ is injective. Let $x \in \mathrm{Ker}(\partial^i[k])$ with $g^i(x) \in \mathrm{Im}(\partial^{i-1}[k+1])$. We need to show that $x \in \mathrm{Im}(\partial^{i-1}[k])$. Let $y = \sum_\sigma y_\sigma \cdot 1_\sigma \in G^{i-1}$ such that $\partial^{i-1}[k+1](y) = g^i(x)$. We can write y_σ uniquely as $y'_\sigma + \mu_\sigma y''_\sigma$ where no monomial in $\mathrm{supp}(y'_\sigma)$ is divided by μ_σ. If

$$y' = \sum_\sigma y'_\sigma \cdot 1_\sigma, \quad y'' = \sum_\sigma y''_\sigma \cdot 1_\sigma,$$

then, since $g^i(x) = \partial^{i-1}[k+1](y)$, we have:

$$g^i(x) = \partial^{i-1}[k+1](y') + \partial^{i-1}[k+1](g^{i-1}(y'')) = \partial^{i-1}[k+1](y') + g^i(\partial^{i-1}[k](y'')). \tag{3.1}$$

Writing $z = \sum_\sigma z_\sigma \cdot 1_\sigma := \partial^{i-1}[k+1](y') \in G^i$, we have

$$z_\sigma = \sum_{v \in \sigma} \mathrm{sign}(v, \sigma) \left(\frac{\mu_\sigma}{\mu_{\sigma \setminus \{v\}}} \right)^{k+1} y'_{\sigma \setminus \{v\}}.$$

Since J is squarefree and $\mu_{\sigma \setminus \{v\}}$ does not divide any monomial in $\mathrm{supp}(y'_{\sigma \setminus \{v\}})$ for any $v \in \sigma$, μ_σ cannot divide z_σ unless the latter is zero. On the other hand, μ_σ must divide the 1_σ-component of $g^i(x)$ and so, exploiting (3.1), z_σ. Therefore $z_\sigma = 0$, and since σ was arbitrary $z = 0$, that is: $g^i(x) = g^i(\partial^{i-1}[k](y''))$. Being $g^i : G^i \to G^i$ obviously injective, we have found $x = \partial^{i-1}[k](y'')$. □

Finally, the main result of these lectures, using Proposition 3.4 and Corollary 3.3, follows by Corollary 3.4:

Corollary 3.5 *Let $I \subset R = K[X_1, \ldots, X_n]$ be a homogeneous ideal and $<$ a monomial order such that $\mathrm{in}(I) \subset R$ is a squarefree monomial ideal. Then*

$$\dim_K (H_\mathfrak{m}^i(R/I)_j) = \dim_K (H_\mathfrak{m}^i(R/\mathrm{in}(I))_j) \ \forall \, i, j \in \mathbb{Z}.$$

In particular $\mathrm{depth} R/\mathrm{in}(I) = \mathrm{depth} R/I$ *and* $\mathrm{reg} R/\mathrm{in}(I) = \mathrm{reg} R/I$.

3.5 Some Discussions and Related Open Questions

We conclude the chapter with some final speculations. In view of Theorem 3.2 and Corollary 3.3 it is interesting to study the fibre-full A-modules, in particular when $A = P = K[X_1, \ldots, X_n, t]$. Although this notion is defined on the A-module M, there are interesting situations in which a nice special fibre M/tM ensures that M is fibre-full. We have already seen an example in Corollary 3.4, where the fact that M/tM is a Stanley-Reisner ring gives the fibre-full property of M when $A = P$.

Dao, De Stefani and Ma introduced in [7] the notion of *cohomologically full* ring. In particular, a quotient of the form P/J where J is a homogeneous ideal of P (with respect to the $(w, 1) = (w_1, \ldots, w_n, 1)$-grading) is cohomologically full if one of the following two equivalent conditions holds:

- For any ideal $H \subset J$ of P such that $\sqrt{H} = \sqrt{J}$, the natural map $\mathrm{Ext}_P^i(P/J, P) \to \mathrm{Ext}_P^i(P/H, P)$ is injective for all $i \in \mathbb{N}$.
- For any inverse system $\{J_k\}_{k \in \mathbb{N} \setminus \{0\}}$ of ideals of P cofinal with $\{J^k\}_{k \in \mathbb{N} \setminus \{0\}}$ (namely, for each $k \in \mathbb{N} \setminus \{0\}$, there exists $h \in \mathbb{N} \setminus \{0\}$ such that $J^k \supset J_h$ and $J_k \supset J^h$), the natural map $\mathrm{Ext}_P^i(P/J, P) \to \mathrm{Ext}_P^i(P/J_k, P)$ is injective for all $i \in \mathbb{N}, k \in \mathbb{N} \setminus \{0\}$.

So, as a consequence of Theorem 3.3, essentially already proved by Lyubeznik in [10], if J is a squarefree monomial ideal, then P/J is cohomologically full. We have:

Proposition 3.6 *For a graded P-module M such that t is a nonzerodivisor, if M/tM is a cohomologically full ring, then M is fibre-full.*

Proof Since t is a nonzerodivisor on M and $M/tM \cong P/J$ for some homogeneous ideal $J \subset P$, then $M \cong P/U$ for some homogeneous ideal $U \subset P$ with $J = U + tP$. In particular, $J^k \subset U + t^k P \subset J$, so the injective map $\mathrm{Ext}_P^i(P/J, P) \to \mathrm{Ext}_P^i(P/J^k, P)$ factors as $\mathrm{Ext}_P^i(P/J, P) \to \mathrm{Ext}_P^i(P/(U + t^k P), P) \to \mathrm{Ext}_P^i(P/J^k, P)$. Hence the first map $\mathrm{Ext}_P^i(M/tM, P) \to \mathrm{Ext}_P^i(M/t^k M, P)$ is injective as well. $\qquad \square$

The "cohomologically full" notion is not limited to Stanley-Reisner rings, but applies to several other classes of rings: for example, it is easy to see that P/J is cohomologically full whenever P/J is Cohen-Macaulay, even though this is not very useful for our purposes). More interestingly, we have (see [7] for more details):

- In positive characteristic, if P/J is F-pure then it is cohomologically full (Ma, [11, Theorem 3.7]).
- In characteristic 0, if P/J is Du Bois then it is cohomologically full (Ma, Schwede and Shimomoto, [13, Lemma 3.3]).

3.5.1 Cohomologically Full Rings Defined by Monomial Ideals

In view of Corollary 3.3 and Proposition 3.6, it would be interesting to study which monomial ideals J of $R = K[X_1, \ldots, X_n]$ are such that R/J is cohomologically full. In fact we have:

Proposition 3.7 Let $I \subset R = K[X_1, \ldots, X_n]$ be a homogeneous ideal such that $R/\mathrm{in}(I)$ is cohomologically full. Then

$$\dim_K (H_{\mathfrak{m}}^i (R/I)_j) = \dim_K (H_{\mathfrak{m}}^i (R/\mathrm{in}(I))_j) \ \forall \, i, j \in \mathbb{Z}.$$

In particular $\mathrm{depth} R/\mathrm{in}(I) = \mathrm{depth} R/I$ and $\mathrm{reg} R/\mathrm{in}(I) = \mathrm{reg} R/I$.

It is easy to check that the same proof as Theorem 3.3 works if J is minimally generated by monomials $\mu_1, \ldots, \mu_r \in R$ such that, for each $i = 1, \ldots, n$, there exists $k_i \in \mathbb{N} \setminus \{0\}$ such that $X_i | \mu_j \iff X_i^{k_i} | \mu_j$ for any $j = 1, \ldots, r$.

Problem 3.1 Characterize the (or find a big class of) monomial ideals $J \subset R = K[X_1, \ldots, X_n]$ such that R/J is cohomologically full.

3.5.2 Sagbi Bases

In this chapter, while focusing on Gröbner bases, we have actually also treated Sagbi bases. Given a K-subalgebra $B \subset R = K[X_1, \ldots, X_n]$, and fixed a monomial order on R, the *initial algebra* of B is defined as:

$$\mathrm{in}(B) := K[\mathrm{in}(f) : f \in B] \subset R.$$

Even if B is finitely generated, the initial algebra $\mathrm{in}(B)$ may not be. However, if it is, we can pick $f_1, \ldots, f_m \in B$ such that

$$\mathrm{in}(B) = K[\mathrm{in}(f_1), \ldots, \mathrm{in}(f_m)].$$

Such $f_1, \ldots, f_m$ are called a *Sagbi basis* of B, and it is easy to see that $B = K[f_1, \ldots, f_m]$. In such a situation, it can be shown that, if $B \cong T/J$ where T

is a polynomial ring in m variables over K, there exists a weight vector $w = (w_1, \ldots, w_m) \in (\mathbb{N} \setminus \{0\})^m$ such that $\mathrm{in}(B) \cong T/\mathrm{in}_w(J)$ (cf. [2, Lemma 2.2]). So, by Corollary 3.3 and Proposition 3.6 we have the following:

Proposition 3.8 *If B is a graded K-subalgebra of $R = K[X_1, \ldots, X_n]$ with a finite Sagbi basis for a monomial order on R such that $\mathrm{in}(B)$ is cohomologically full, then, if $\mathrm{m} = (X_1, \ldots, X_n)$,*

$$\dim_K (H^i_{\mathrm{m} \cap B}(B)_j) = \dim_K (H^i_{\mathrm{m} \cap \mathrm{in}(B)}(\mathrm{in}(B))_j) \ \forall \, i, j \in \mathbb{Z}.$$

In particular $\mathrm{depth}(\mathrm{in}(B)) = \mathrm{depth} B$ *and* $\mathrm{reg}(\mathrm{in}(B)) = \mathrm{reg} B$.

In view of the Proposition above, we propose the following problem:

Problem 3.2 Characterize the (or find a big class of) finitely generated monoids $M \subset \mathbb{Z}^n$ such that $K[M] := K[\mathbf{X^u} : \mathbf{u} \in M]$ is cohomologically full.

It should be noticed that, in characteristic 0, a class of finitely generated monoids $M \subset \mathbb{Z}^n$ such that $K[M]$ is cohomologically full consists of all *seminormal* *monoids* (namely, if $2v, 3v \in M$ for some $v \in \mathbb{Z}^n$, then $v \in M$; equivalently, $K[M]$ is a seminormal ring). In fact, if M is seminormal, then $K[M]$ is F-pure whenever K is a field of characteristic $p \gg 0$ (cf. [5, Propositions 6.1 and 6.2]); in particular, if K has characteristic 0 and M is seminormal, then $K[M]$ is Du Bois by a result of Schwede [16, Theorem 6.1], and so cohomologically full. Summarizing we have:

Proposition 3.9 *If B is a graded K-subalgebra of $R = K[X_1, \ldots, X_n]$ with a finite Sagbi basis for a monomial order on R such that $\mathrm{in}(B)$ is seminormal, then, if $\mathrm{m} = (X_1, \ldots, X_n)$,*

$$\dim_K (H^i_{\mathrm{m} \cap B}(B)_j) = \dim_K (H^i_{\mathrm{m} \cap \mathrm{in}(B)}(\mathrm{in}(B))_j) \ \forall \, i, j \in \mathbb{Z}.$$

In particular $\mathrm{depth} \, \mathrm{in}(B) = \mathrm{depth} B$ *and* $\mathrm{reg} \, \mathrm{in}(B) = \mathrm{reg} B$.

3.5.3 The Category of Fibre-Full Modules

While the notion of cohomologically full rings is very powerful, it is not immediately clear how to extend this concept to modules in the right way: this is the main reason why we introduced the concept of fibre-full modules. For example, adopting the same notation as Sect. 3.3, it would be very interesting to answer the following:

Problem 3.3 Let M be a fibre-full A-module and suppose that $A = \oplus_{i \in \mathbb{N}} A_i$ is a graded Gorenstein ring with $A_0 = K[t]$. Is the canonical module of M a fibre-full A-module? More generally, is $\mathrm{Ext}^i_A(M, A)$ a fibre-full A-module for all $i \in \mathbb{N}$?

An affirmative answer to the question above would have consequences on the preservation of the canonical Cohen-Macaulay property (concept introduced by

Schenzel [15]) and of the homological degree (concept introduced by Vasconcelos [17]) along deformations. As a more general problem, I would like to mention the following:

Problem 3.4 Study the full subcategory of the fibre-full A-modules (in the category of A-modules) and, where possible, compare it with the category of squarefree modules (see Yanagawa [18, 19]).

References

1. D. Bayer, I. Peeva, B. Sturmfels, *Monomial Resolutions*. Mathematical Research Letters, vol. 5, pp. 31–46 (1998)
2. W. Bruns, A. Conca, *Gröbner Bases and Determinantal Ideals*. Commutative Algebra, Singularities and Computer Algebra, pp. 9–66 (Sinaia, 2002). NATO Sci. Ser. II Math. Phys. Chem., vol. 115 (Kluwer Acad. Publ., Dordrecht, 2003)
3. W. Bruns, J. Gubeladze, *Polytopes. Rings and K-Theory*, Springer Monographs in Mathematics (Springer, Dordecht, 2009)
4. W. Bruns, J. Herzog, *Cohen-Macaulay Rings*. Cambridge Studies in Advanced Mathematics, vol. 39 (Cambridge University Press, Cambridge, 1993)
5. W. Bruns, P. Li, T. Roemer, On seminormal monoid rings. J. Algebra **302**, 361–386 (2006)
6. A. Conca, M. Varbaro, Square-free Gröbner degenerations. Invent. Math. **221**(3), 713–730 (2020)
7. H. Dao, A. De Stefani, L. Ma, *Cohomologically Full Rings*. IMRN, https://doi.org/10.1093/imrn/rnz203. arXiv:1806.00536
8. D. Eisenbud, *Commutative Algebra with a View Toward Algebraic Geometry*. Graduate Texts in Mathematics, vol. 150 (Springer, 1994)
9. J. Kollár, S.J. Kovács, Deformations of Log Canonical and F-pure singularities. Algebr. Geom. **7**(6), 758–780 (2020)
10. G. Lyubeznik, *On the local cohomology modules $H_a^i(R)$ for ideals a generated by monomials in an R-sequence*, in Complete Intersections (Acireale, 1983), vol. 1092. Lecture Notes in Math. (Springer, Berlin, 1984), pp. 214–220
11. L. Ma, Finiteness properties of local cohomology for F-pure local rings. Int. Math. Res. Not. IMRN **20**, 5489–5509 (2014)
12. L. Ma, H.P. Quy, Frobenius actions on local cohomology modules and deformation. Nagoya Math. J. **232**, 55–75 (2018)
13. L. Ma, K. Schwede, K. Shimomoto, Local cohomology of Du Bois singularities and applications to families. Compos. Math. **153**, 2147–2170 (2017)
14. H. Matsumura, *Commutative Ring Theory*. Cambridge Studies in Advanced Mathematics, vol. 8 (Cambridge University Press, Cambridge, 1986)
15. P. Schenzel, On birational Macaulayfications and Cohen-Macaulay canonical modules. J. Algebra **275**, 751–770 (2004)
16. K. Schwede, F-injective singularities are Du Bois. Am. J. Math. **131**, 445–473 (2009)
17. W. Vasconcelos, The homological degree of a module. Trans. Am. Math. Soc. **350**, 1167–1179 (1998)
18. K. Yanagawa, *Squarefree Modules and Local Cohomology Modules at Monomial Ideals*. Local Cohomology and Its Applications (Guanajuato, 1999). Lecture Notes in Pure and Appl. Math., Vol. 226 (Dekker, New York, 2002), pp. 207–231
19. K. Yanagawa, Derived category of squarefree modules and local cohomology with monomial ideal support. J. Math. Soc. Jpn. **56**, 289–308 (2004)

Chapter 4
Adams Operations in Commutative Algebra

Mark E. Walker

Abstract The classical *Adams operations* are defined on the Grothendieck group of algebraic vector bundles on a scheme, and are related to Grothendieck's Riemann-Roch Theorem. In these notes, I review the theory of Adams operations for Grothendieck groups with supports, following the work of Gillet and Soulé, and give two applications in local algebra: A proof of the Serre Vanishing Conjecture for arbitrary regular local rings (due to Gillet and Soulé) and a proof of the Total Rank Conjecture for many local rings.

4.1 Classical Lambda and Adams Operations on K_0

In this section, I outline the basic definitions and properties concerning lambda and Adams operations on classical Grothendieck groups. The lambda operations were developed by Grothendieck as a tool for his proof of the Riemann-Roch Theorem [16]. The related Adams operations were first introduced in algebraic topology by Frank Adams [1, Theorem 4.1]; see also [2, §2]. A more systematic and thorough treatment of the material in this section may be found in the book by Fulton and Lang [8].

Unless otherwise stated, all rings are assumed to be commutative and noetherian and all schemes are assumed to be separated and noetherian.

M. E. Walker (✉)
University of Nebraska-Lincoln, Lincoln, NE, USA
e-mail: mark.walker@unl.edu

© The Editor(s) (if applicable) and The Author(s), under exclusive license
to Springer Nature Switzerland AG 2021
A. Conca et al. (eds.), *Recent Developments in Commutative Algebra*, Lecture
Notes in Mathematics 2283, https://doi.org/10.1007/978-3-030-65064-3_4

4.1.1 Definition of K_0

For a (commutative, noetherian) ring R, let $P_{fg}(R)$ denote the category of all finitely generated and projective R-modules. (If one is worried about set-theoretic concerns that arise in what follows, one can take $P_{fg}(R)$ to be those modules that are direct summands of R^N for some N.)

Definition 4.1 For a (commutative, noetherian) ring R, define $K_0(R)$ to be the abelian group generated by classes $[P]$ of objects $P \in \mathrm{ob}\, P_{fg}(R)$ with relations coming from short exact sequences of such: For each short exact sequence $0 \to P' \to P \to P'' \to 0$, we declare

$$[P] = [P'] + [P''].$$

Remark 4.1 Since each such short exact sequence is split exact, one may equivalently define $K_0(R)$ as the group completion of the additive monoid of isomorphism classes of $P_{fg}(R)$, with addition given by direct sum.

For a pair of finitely generated projective R-modules P_1, P_2, the split exact sequence $0 \to P_1 \to P_1 \oplus P_2 \to P_2 \to 0$ implies that $[P_1] + [P_2] = [P_1 \oplus P_2]$ holds in $K_0(R)$. In particular, every element of $K_0(R)$ can be written as a formal difference $[P] - [P']$.

More generally, when X is a (noetherian, separated) scheme, $K_0(X)$ is the abelian group generated by classes of locally free coherent sheaves on X, modulo relations coming from short exact sequences of such. Nearly everything in this section concerning K_0 of rings generalizes to schemes, but I will sometimes leave such generalizations unspoken.

Given an abelian group A, a function $\rho \colon \mathrm{ob}\, P_{fg}(R) \to A$ is called *additive on short exact sequences* (or just *additive*) if $\rho(P) = \rho(P') + \rho(P'')$ holds whenever there is a short exact sequence $0 \to P' \to P \to P'' \to 0$. For any abelian group A, there is a bijection of sets

$$\mathrm{Hom}_{Ab}(K_0(R), A)) \cong \{\text{additive functions } \rho \colon \mathrm{ob}\, P_{fg}(R) \to A\}.$$

Example 4.1 If $\mathrm{Spec}(R)$ is connected, each object P of $P_{fg}(R)$ has a well-defined rank, written $\mathrm{rank}(P) \in \mathbb{Z}$, which may be defined as the rank of the free module obtained by localizing at any point of $\mathrm{Spec}(R)$. The rank function is additive and hence induces a homomorphism

$$\mathrm{rank} \colon K_0(R) \to \mathbb{Z}$$

that sends a typical element $[P_1] - [P_2]$ of $K_0(R)$ to $\mathrm{rank}(P_1) - \mathrm{rank}(P_2)$.

If R is a local ring, then since every projective module is free, the rank function induces an isomorphism

$$\text{rank}: K_0(R) \xrightarrow{\cong} \mathbb{Z}.$$

If R is a Dedekind domain (a regular domain of dimension 1), then the kernel of rank: $K_0(R) \to \mathbb{Z}$ is isomorphic to the divisor class group of R (and also to the Picard group of R). For other rings, the kernel is more difficult to describe.

The assignment $R \mapsto K_0(R)$ is functorial: Given a ring homomorphism $g: R \to S$, let

$$g_*: K_0(R) \to K_0(S)$$

be the map induced by extension of scalars: given $P \in P_{\text{fg}}(R)$, we have $g_*([P]) = [P \otimes_R S]$. The map g_* is well-defined since the function $P \mapsto [P \otimes_R S]$ is additive on short exact sequences.

The abelian group $K_0(R)$ becomes a commutative ring under the operation induced by tensor product:

$$[P] \cdot [P'] := [P \otimes_R P'].$$

This is a well-defined operation since for a fixed P, the mapping $P' \mapsto [P \otimes_R P']$ is additive on short exact sequences, and similarly for $P \mapsto [P \otimes_R P']$ for each fixed P'. The identity element is $[R]$. Moreover, this structure is natural, so that $K_0(-)$ is a covariant functor from commutative rings to commutative rings.

The same holds on the level of schemes, but the variance is of course opposite: $K_0(-)$ is a contravariant functor from noetherian schemes to commutative rings.

4.1.2 Lambda Operations

An important additional structure on $K_0(R)$ arises from the exterior power functors. For $P \in \text{ob } P_{\text{fg}}(R)$, let $\Lambda_R^k(P)$ (or sometimes just $\Lambda^k(P)$) denote the k-th exterior power, defined as

$$\Lambda_R^k(P) = \frac{\overbrace{P \otimes_R \cdots \otimes_R P}^{k}}{L}$$

where L is the sub-module generated by

$$x_1 \otimes \cdots \otimes x_k - \text{sign}(\sigma) x_{\sigma(1)} \otimes \cdots \otimes x_{\sigma(k)}$$

for all elements $x_1, \ldots, x_k \in P$ and permutations $\sigma \in S_k$. The image of $x_1 \otimes \cdots \otimes x_k$ in $\Lambda_R^k(P)$ is written as $x_1 \wedge \cdots \wedge x_k$.

Since the k-th exterior power functor localizes well and the k-the exterior power of a free module of rank r is free of rank $\binom{r}{k}$, it follows that $\Lambda_R^k(P)$ is a projective module of rank equal to $\binom{\mathrm{rank}(P)}{k}$ (assuming $\mathrm{Spec}(R)$ is connected). In particular, $\Lambda_R^k(P) = 0$ for $k > \mathrm{rank}(P)$.

It is important to notice that the exterior power functor is *not additive* on short exact sequences. However, there is a replacement for this lack of additivity: given a short exact sequence

$$0 \to P' \xrightarrow{\iota} P \xrightarrow{\pi} P'' \to 0 \tag{4.1}$$

there is a filtration

$$0 = F_{-1} \subseteq F_0 \subseteq \cdots \subseteq F_k = \Lambda^k(P)$$

such that F_i/F_{i-1} is isomorphic to $\Lambda^{k-i}(P') \otimes_R \Lambda^i(P'')$, for each $0 \le i \le k$. Namely, define F_i as the image of the map $(P')^{\otimes k-i} \otimes_R P^{\otimes i} \to \Lambda^k(P)$ given on generators by

$$x_1' \otimes \cdots \otimes x_{k-i}' \otimes y_1 \otimes \cdots \otimes y_i \mapsto \iota(x_1') \wedge \cdots \wedge \iota(x_{k-i}') \wedge y_1 \wedge \cdots \wedge y_i.$$

See [10, §2].

For example, when $k = 2$, we have the filtration

$$0 \subseteq \Lambda^2(P') \subseteq F_1 \subseteq \Lambda^2(P)$$

with

$$F_1 = \mathrm{im}(P' \otimes P \xrightarrow{x' \otimes y \mapsto \iota(x') \wedge y} \Lambda^2(P))$$

The isomorphism $F_1/\Lambda^2(P') \xrightarrow{\cong} P' \otimes P''$ is induced from the surjection $\mathrm{id} \otimes \pi : P' \otimes_R P \to P' \otimes_R P''$, and the isomorphism $\Lambda^2(P)/F_1 \cong \Lambda^2(P'')$ is induced by the surjection $\Lambda^2(\pi) : \Lambda^2(P) \twoheadrightarrow \Lambda^2(P'')$.

In general, from this filtration we obtain the equation

$$[\Lambda^k(P)] = \sum_{i=0}^{k} [\Lambda^{k-i}(P')] \cdot [\Lambda^i(P'')] \tag{4.2}$$

in $K_0(R)$. For example

$$[\Lambda^2(P)] = [\Lambda^2(P')] + [P'][P''] + [\Lambda^2(P'')].$$

Lemma 4.1 (Grothendieck) *For each commutative ring R, there are functions $\lambda^k \colon K_0(R) \to K_0(R)$, for $k \geq 0$, uniquely determined by the following properties:*

- *For each $P \in P_{\mathrm{fg}}(R)$ and k, we have*

$$\lambda^k([P]) = [\Lambda^k(P)].$$

- *For all $\alpha, \beta \in K_0(R)$ and all k, we have*

$$\lambda^k(\alpha + \beta) = \sum_{i=0}^{k} \lambda^i(\alpha)\lambda^{k-i}(\beta). \tag{4.3}$$

Moreover, these operators are natural for ring maps.

Proof Let t be a formal parameter, form the ring $K_0(R)[[t]]$ of power series with coefficients in the commutative ring $K_0(R)$, and for any $P \in P_{\mathrm{fg}}(R)$, define

$$\lambda_t(P) := \sum_k [\Lambda^k(P)]t^k \in K_0(R)[[t]].$$

(The sum is actually finite since $\Lambda^k(P) = 0$ for $k > \mathrm{rank}(P)$.)

Since the constant term of $\lambda_t(P)$ is 1, $\lambda_r(P)$ belongs to the group of units of $K_0(R)[[t]]$. For each short exact sequence (4.1), it follows from (4.2) that

$$\lambda_t(P) = \lambda_t(P') \cdot \lambda_t(P'');$$

that is, λ_t is additive on short exact sequences, provided we interpret it as taking values in the multiplicative abelian group $K_0(R)[[t]]^\times$. Therefore, by the universal mapping property of the Grothendieck group, λ_t induces a homomorphism

$$\lambda_t \colon K_0(R) \to K_0(R)[[t]]^\times$$

of abelian groups. Finally, we define

$$\lambda^k \colon K_0(R) \to K_0(R)$$

to be the composition of λ_t with the map $K_0(R)[[t]]^\times \to K_0(R)$ sending a power series to the coefficient of t^k. Equation (4.3) follows.

The uniqueness property is seen to hold by induction on k.

The naturality assertion follows from the fact that given a ring map $R \to S$, we have an isomorphism $\lambda_R^k(P) \otimes_P S \cong \lambda_S^k(P \otimes_R S)$. □

The operator λ^0 is the constant function with value $1 = [R] \in K_0(R)$, and λ^1 is the identity map. The operator λ^k is *not* a homomorphism of abelian groups for $k \neq 1$.

Example 4.2 What is $\lambda^2(-[P])$? We have

$$
\begin{aligned}
0 &= \lambda^2([P] + (-[P])) \\
&= \lambda^2([P])\lambda^0(-[P]) + \lambda^1([P])\lambda^1(-[P]) + \lambda^0([P])\lambda^2(-[P]) \\
&= [\Lambda^2(P)] - [P \otimes P] + \lambda^2(-[P])
\end{aligned}
$$

and hence

$$
\lambda^2(-[P]) = [P \otimes P] - [\Lambda^2(P))].
$$

Example 4.3 When R is local, we have $K_0(R) \cong \mathbb{Z}$. Under this isomorphism the operator λ^k corresponds to the operator on $\mathbb{Z}$, which we will also write at λ^k, that satisfies

$$
\lambda^k(n) = \binom{n}{k} = \frac{n(n-1)\cdots(n-k+1)}{k!}
$$

at least when $n \geq 0$. This holds since $\Lambda^k(R^n)$ is free of rank $\binom{n}{k}$ for any $n \geq 0$. But what if $n < 0$? To figure this out, we use the notation and results of the proof of the lemma. We have

$$
\lambda_t(-m) = 1/\lambda_t(m) \in \mathbb{Z}[[t]]
$$

and so if $m > 0$, we get

$$
\lambda_t(-m) = \frac{1}{1 + mt + \binom{m}{2}t^2 + \cdots} = \frac{1}{(1+t)^m} = (1 - t + t^2 - t^3 + \cdots)^m.
$$

Thus $\lambda^k(-m)$ is the coefficient of t^m in $(1 - t + t^2 - t^3 + \cdots)^m$. This shows

$$
\lambda^k(n) = \frac{n(n-1)\cdots(n-k+1)}{k!}
$$

is the correct formula even when $n < 0$.

The commutative ring $K_0(R)$ equipped with the operators $\{\lambda^k\}_{k \geq 0}$, form what's called a *(special) lambda ring*. Roughly, this means that

$$
\lambda^k(a + b) = \sum_{i=0}^{k} \lambda^{k-i}(a)\lambda^i(b),
$$

for all a, b and that the rules describing how the operators interact with multiplication and composition are given by certain universal polynomials. In detail, we have

$$\lambda^k(a \cdot b) = P_k(\lambda^1(a), \ldots, \lambda^k(a), \lambda^1(b), \ldots, \lambda^k(b))$$

for some polynomial P_k in $2k$ variables, and similarly for $\lambda^k(\lambda^j(a))$. I will not need the details.

More generally, starting with the exterior power functors for locally free coherent sheaves, one defines lambda operators on $K_0(X)$, making it into a special lambda ring. Even if one is only interested in Grothendieck groups of commutative rings, passing to schemes is valuable due to the following important fact:

Lemma 4.2 (The Splitting Principle) *For each noetherian ring R and $P \in P_{\mathrm{fg}}(R)$, there exists a morphism of noetherian schemes $p : X \to \mathrm{Spec}(R)$ such that*

1. *the induced map $p^* : K_0(R) \to K_0(X)$ is injective and*
2. *$p^*[P] = \sum_{i=1}^{\mathrm{rank}(P)}[L_i]$ where the L_i's are line bundles on X (i.e., coherent sheaves that are locally free of rank 1).*

See [8, Theorem V.2.3] for a proof.

More generally, the analogous result holds for schemes: starting with any noetherian scheme Y and locally free coherent sheaf P on Y, there is a morphism $p : X \to Y$ such that the above two properties hold. In fact, X may be taken to be the flag variety over Y associated to P.

With the notation of the Splitting Principle, since $\Lambda^k(L_i) = 0$ for all $k \geq 2$, we have

$$p^*\lambda_t([P]) = \lambda_t(p^*[P]) = \prod_i \lambda_t([L_i]) = \prod_i (1 + [L_i]t)$$

and hence

$$p^*\lambda^k([P]) = \sum_{i_1 < \cdots < i_k} [L_{i_1}] \cdots [L_{i_k}]$$

for any k. The expression on right side of this equation belongs, a priori, to $K_0(X)$, but since the left hand side is in the image of p^*, we may interpret this equation as occurring in $K_0(R)$ (but not so for the individual terms of the right-hand side).

We may reinterpret this as saying that $\lambda^k([P])$ is the k-th elementary symmetric polynomial evaluated on $[L_1], \ldots, [L_k]$. In detail, recall that for each non-negative integer k and variables $x_1, \ldots, x_r$, the k-th elementary symmetric polynomial $\sigma_k(x_1, \ldots, x_n)$ is the coefficient of t^k in the expansion of $\prod_{i=1}^r (1 + x_i t)$. For example, $\sigma_1 = \sum_i x_i$ and $\sigma_2 = \sum_{i<j} x_i x_j$. Then, abusing notation slightly, we have

$$\lambda^k([P]) = \sigma_k([L_1], \ldots, [L_r])$$

where $r = \mathrm{rank}(P)$ and the L_i's are the "formal line components" of P.

4.1.3 Adams Operations

The Adams operations are defined by combining the lambda operations together in a specific way so as to produce operators that are linear. They are based on work of Frank Adams [1], working in the context of topological vector bundles.

The k-th Adams operation ψ^k on $K_0(R)$ is defined as follows. Recall the homomorphism of abelian groups $\lambda_t \colon K_0(R) \to K_0(R)[[t]]^\times$ defined in the proof of Lemma 4.1. For any commutative ring A and variable t, write dlog: $A[[t]]^\times \to A[[t]]$ for the function sending $p(t) = \sum_i a_i t^i$ to

$$\mathrm{dlog}(p(t)) := \frac{p'(t)}{p(t)} = \frac{\sum_i i a_i t^{i-1}}{\sum_i a_i t^i}$$

("the derivative of the natural log of $p(t)$"). Then dlog converts multiplication to addition; i.e. it is a homomorphism of abelian groups, where the operation on the source is power series multiplication and the operation on the target is addition of power series.

For $\alpha \in K_0(R)$, set

$$\psi_t(\alpha) = \mathrm{rank}(\alpha) - t\,\mathrm{dlog}(\lambda_{-t}(\alpha)). \tag{4.4}$$

Since dlog converts multiplication to addition, $\psi_t \colon K_0(R) \to K_0(R)[[t]]$ is a homomorphism of additive groups. We define

$$\psi^k \colon K_0(R) \to K_0(R)$$

as the composition of ψ_t with the map that sends a power series to its t^k coefficient. By construction, ψ^k is a homomorphism of abelian groups for each $k \geq 0$.

Example 4.4 Let's compute ψ^k for $k \leq 3$. We have

$$
\begin{aligned}
\mathrm{dlog}(\lambda_{-t}(\alpha)) &= \frac{\lambda_{-t}(\alpha)'}{\lambda_{-t}(\alpha)} \\
&= \frac{-\lambda^1(\alpha) + 2\lambda^2(\alpha)t - 3\lambda^3(\alpha)t^2 + \cdots}{1 - \lambda^1(\alpha)t + \lambda^2(\alpha)t^2 - \cdots} \\
&= (-\lambda^1(\alpha) + 2\lambda^2(\alpha)t - \cdots)(1 + \lambda^1(\alpha)t + (\lambda^1(\alpha)^2 - \lambda^2(\alpha))t^2 + \cdots) \\
&= -\lambda^1(\alpha) + (-\lambda^1(\alpha)^2 + 2\lambda^2(\alpha))t + \\
&\quad (-\lambda^1(\alpha)^3 + \lambda^1(\alpha)\lambda^2(\alpha) + 2\lambda^2(\alpha)\lambda^1(\alpha) - 3\lambda^3(\alpha)^2)t^2 + \cdots.
\end{aligned}
$$

Since $\lambda^1(\alpha) = \alpha$, it follows that

$$\psi^0(\alpha) = \text{rank}(\alpha)$$

$$\psi^1(\alpha) = \alpha$$

$$\psi^2(\alpha) = \alpha^2 - 2\lambda^2(\alpha)$$

$$\psi^3(\alpha) = \alpha^3 - 3\lambda^2(\alpha)\alpha + 3\lambda^3(\alpha)$$

Let us double check that ψ^2 is additive on short exact sequences directly, using the formula established in the example: Given $0 \to P' \to P \to P'' \to 0$, we have

$$
\begin{aligned}
\psi^2(P) &= [P]^2 - 2[\Lambda^2(P)] \\
&= ([P'] + [P''])^2 - 2([\Lambda^2(P')] + [P' \otimes P''] + [\Lambda^2(P'')]) \\
&= [P']^2 - 2[\Lambda^2(P')] + [P'']^2 - 2[\Lambda^2(P'')] \\
&= \psi^2(P') + \psi^2(P'').
\end{aligned}
$$

In general, Eq. (4.4) leads to the recursive formula

$$\psi^k(\alpha) - \psi^{k-1}(\alpha)\lambda^1(\alpha) + \psi^{k-2}(\alpha)\lambda^2(\alpha) - \cdots + (-1)^{k-1}\psi^1(\alpha)\lambda^{k-1}(\alpha) + (-1)^k k\lambda^k(\alpha) = 0. \tag{4.5}$$

The key properties of the Adams operations are summarized in:

Proposition 4.1 *For each $k \geq 0$,*

1. *$\psi^k \colon K_0(R) \to K_0(R)$ is a ring endomorphism for each R (and likewise for $K_0(X)$ for each X),*
2. *ψ^k is natural for ring maps (and more generally for morphisms of schemes),*
3. *if L is a rank one projective R-module (or, more generally, a coherent sheaf locally free of rank 1 on a scheme), then $\psi^k([L]) = [L^{\otimes k}]$.*

Moreover, these properties uniquely characterize the operator ψ^k.

Proof We have already observed that ψ^k preserves addition.

Property 2 follows by induction on k using (4.5) and the fact that the lambda operations are natural.

Property 3 also follows from (4.5) by induction on k, using also that $\lambda^j([L]) = 0$ for all $j \geq 2$.

Multiplicativity follows from additivity and the Splitting Principle. In detail, it suffices to check $\psi^k([P][P']) = \psi^k([P])\psi^k([P'])$. In the case when $[P]$ and $[P']$ are sums of classes of line bundles, this holds by (3) and the fact that ψ^k is additive. The general case follows by naturality and the Splitting Principle.

The last assertion also follows from the Splitting Principle. $\square$

Using this proposition, one may also relate the Adams operations to the lambda operations using the Splitting Principle. Given $P \in P_{\mathrm{fg}}(R)$, let $\pi : X \to \mathrm{Spec}(R)$ be as in the Splitting Principle, so that $\pi^*([P]) = \sum_{i=1}^{r}[L_i]$, where the L_i's are line bundles. Then

$$\pi^* \psi^k([P]) = \sum_{i=1}^{r}[L_i^{\otimes k}] = \sum_{i=1}^{r}[L_i]^k$$

or in other words

$$\psi^k([P]) = N_k([L_1], \ldots, [L_r])$$

where N_k is the *k-th Newton polynomial*, defined by $N_k(x_1, \ldots, x_r) = \sum_i x_i^k$. Since N_k is a symmetric polynomial and every symmetric polynomial can be written in terms of the elementary symmetric polynomials $\sigma_1, \ldots, \sigma_r$, for a given k, there is a polynomial g_k in k variables such that

$$N_k(x_1, \ldots, x_r) = g_k(\sigma_1(x_1, \ldots, x_r), \ldots, \sigma_k(x_1, \ldots, x_r))$$

holds for any r. (This formula is independent of r in the sense that upon setting $x_r = 0$ on both sides, we obtain $N_k(x_1, \ldots, x_{r-1}) = g_k(\sigma_1(x_1, \ldots, x_{r-1}), \ldots, \sigma_k(x_1, \ldots, x_{r-1}))$.) It follows that

$$\psi^k([P]) = g_k(\lambda^1[P], \ldots, \lambda^k[P]).$$

Since both sides are additive, this extends to arbitrary elements of $K_0(R)$.

For example, setting $k = 2$, for any r we have

$$N_2(x_1, \ldots, x_r) = \sum_i x_i^2 = \left(\sum_i x_i\right)^2 - 2\sum_{i<j} x_i x_j = \sigma_1(x_1, \ldots, x_r)^2 - 2\sigma_2(x_1, \ldots, x_r)$$

so that

$$g_2(y_1, y_2) = y_1^2 - 2y_2$$

and thus

$$\psi^2(\alpha) = \alpha^2 - 2\lambda^2(\alpha),$$

for any $\alpha \in K_0(R)$, as previously noted.

Corollary 4.1 *If R has characteristic p for some prime p, then $\psi^p : K_0(R) \to K_0(R)$ coincides with the map induced by extension of scalars along the Frobenius map $\phi : R \to R$.*

Proof The map $\phi_*\colon K_0(R) \to K_0(R)$ is a natural ring homomorphism and one can check that $\phi_*[L] \cong [L^{\otimes p}]$ for a rank one projective R-module. The result follows from the Proposition.   □

Remark 4.2 It follows from Proposition 4.1 that

$$\psi^k \circ \psi^j = \psi^{kj} \tag{4.6}$$

for all $k, j \geq 0$. In particular, this shows ψ^k and ψ^j commute.

4.1.4   A Theorem of Grothendieck

Finally, I mention a famous theorem of Grothendieck. (Technically, Grothendieck did not phrase this Theorem in terms of Adams operations.)

Theorem 4.1 (Grothendieck, 1950s) *If X is a smooth variety over a field, then for any integer $k \geq 2$, the action of ψ^k on $K_0(X)_{\mathbb{Q}}$ is diagonalizable with eigenvalues in the set $k^0, \ldots, k^{\dim(X)}$. In other words there is an internal direct sum decomposition*

$$K_0(X)_{\mathbb{Q}} = \bigoplus_{j=0}^{\dim(X)} K_0(X)^{(j)}$$

where we define $K_0(X)^{(j)} := \ker((\psi^k - k^j)\colon K_0(X)_{\mathbb{Q}} \to K_0(X)_{\mathbb{Q}})$, the eigenspace of the operator $\psi_{\mathbb{Q}}^k$ of eigenvalue k^j.

 Moreover, for each j, we have isomorphisms

$$K_0(X)^{(j)} \cong CH^j(X)_{\mathbb{Q}}$$

where $CH^j(X)$ is the Chow group of codimension j cycles modulo rational equivalence.

Remark 4.3 The subspace $K_0(X)^{(j)}$ of $K_0(X)_{\mathbb{Q}}$ is independent of the choice of $k \geq 2$.

4.2   Algebraic K-Theory with Supports

Much of the material in this section is based on Gillet and Soulé's paper [9]. The material in this section would work just as well for arbitrary noetherian schemes, but for the sake of concreteness I will stick to affine ones. Let us fix some notation:

- R is a commutative noetherian ring.

- Z is a Zariski closed subset of Spec(R). So,

$$Z = V(I) := \{\mathfrak{p} \in \mathrm{Spec}(R) \mid I \subseteq \mathfrak{p}\}$$

  for some ideal $I \subseteq R$.
- $\mathcal{P}^Z(R)$ is the category of bounded complexes of finitely generated and projective R-modules with homology supported on Z. That is, a typical object is a complex of the form

$$P := (\cdots \to 0 \to P_m \to \cdots \to P_n \to 0 \to \cdots)$$

  with each P_i a finitely generated and projective R-module, such that $P_{\mathfrak{p}}$ is an exact complex for all $\mathfrak{p} \in \mathrm{Spec}(R) \setminus Z$. Morphisms in $\mathcal{P}^Z(R)$ are chain maps.

4.2.1 Grothendieck Group with Supports

Definition 4.2 $K_0^Z(R)$, the *Grothendieck group* of $\mathcal{P}^Z(R)$, is the abelian group generated by the set of classes $[P]$ for each object P of $\mathcal{P}^Z(R)$, subject to two types of relations:

$[P] = 0$ if P is exact, and

$[P] = [P'] + [P'']$ if there exists a short exact sequence of complexes of the form

$$0 \to P' \to P \to P'' \to 0.$$

It is not hard to see that we have the following facts:

1. For any pair of objects $P, P' \in \mathcal{P}^Z(R)$,

$$[P] + [P'] = [P \oplus P'] \in K_0^Z(R). \tag{4.7}$$

2. For any $P \in \mathcal{P}^Z(R)$, we have $[\Sigma P] = -[P]$ in K_0^Z, where ΣP is the suspension (shift) of P. (This follows from the short exact sequence $0 \to P \to \mathrm{cone}(\mathrm{id}_P) \to \Sigma P \to 0$.)

3. Given $P, P' \in \mathcal{P}^Z(R)$, if there exists a quasi-isomorphism $\alpha : P \to P'$ in $\mathcal{P}^Z(R)$, then $[P] = [P']$ in $K_0^Z(R)$. (This follows from the short exact sequence $0 \to P' \to \mathrm{cone}(\alpha) \to P \to 0$.)

4. Every element of $K_0^Z(R)$ is equal to one of the form $[P]$ for some $P \in \mathcal{P}^Z(R)$.

Example 4.5 Suppose M is a finitely generated R-module of finite projective dimension, and choose a bounded resolution $P \xrightarrow{\sim} M$ by finitely generated projective modules. Then P is an object of $\mathcal{P}^{\mathrm{supp}(M)}(X)$ and hence determines a class $[P]$ in $K_0^{\text{supp}(M)}(R)$. If P' is another such resolution, then P and P' are homotopy equivalent and hence $[P] = [P']$. So M determines a well-defined element of $K_0^{\text{supp}(M)}(R)$.

Remark 4.4 Associated to $\mathcal{P}^Z(X)$ we have its *homotopy category*, written $\text{hot}(\mathcal{P}^Z(X))$. The objects of this category are the same as for $\mathcal{P}^Z(X)$, but morphisms are chain homotopy equivalence classes of chain maps. The category $\mathcal{P}^Z(X)$ has the structure of a triangulated category. The suspension functor Σ is the usual shift functor for chain complexes. By definition, a triangle $P_1 \to P_2 \to P_3 \to \Sigma P_1$ is distinguished if it is isomorphic (in $\text{hot}(\mathcal{P}^Z(X))$) to one of the form $P \xrightarrow{\alpha} P' \xrightarrow{\text{can}} \text{cone}(\alpha) \xrightarrow{\text{can}} \Sigma P$.

$K_0^Z(R)$ may be equivalently defined as the Grothendieck group of this triangulated category. This means that it is the abelian group generated by objects of $\mathcal{P}^Z(X)$ modulo the two relations

$$[\Sigma P] = -[P]$$

and

$$[P] = [P'] + [P'']$$

whenever there is a distinguished triangle $P' \to P \to P'' \to \Sigma(P')$ in $\text{hot}(\mathcal{P}^Z(X))$.

Example 4.6 There is a canonical isomorphism $K_0^{\text{Spec}\,R}(R) \xrightarrow{\cong} K_0(R)$, that takes the class of an object $P \in \mathcal{P}^{\text{Spec}(R)}(R)$ to the alternating sum of its components. The inverse takes the class of a finitely generated projective R-module to the class of the complex obtained by regarding it as a complex concentrated in degree 0.

Using the universal mapping property for presentations of abelian groups, we have the following: If A is any abelian group and ρ is a function assigning to each object of $\mathcal{P}^Z(X)$ an element of A such that

$$\rho(P) = 0, \text{ if } P \text{ is acyclic}$$

and

$$\rho(P) = \rho(P') + \rho(P''), \text{ if there is a short exact sequence } 0 \to P' \to P \to P'' \to 0,$$

then ρ induces a unique homomorphism of abelian groups (also written as ρ)

$$\rho : K_0^Z(X) \to A,$$

such that $\rho([P]) = \rho(P)$.

4.2.2 Complexes with Finite Length Homology

Of particular interest for us will be the category $\mathcal{P}^{\mathfrak{m}}(R)$ and its associated Grothendieck group $K_0^{\mathfrak{m}}(R)$, when $(R, \mathfrak{m})$ is a local ring. (Technically the superscripts should be "{m}" not "m", but I will use the latter.) We may equivalently describe an object of $\mathcal{P}^{\mathfrak{m}}(R)$ as a bounded complex of finitely generated projective R-modules having finite length homology. In particular, if M is a finite length R-module of finite projective dimension, then M determines a class in $K_0^{\mathfrak{m}}(R)$ by choosing a free resolution.

Lemma 4.3 *For any ring R and maximal ideal $\mathfrak{m}$ the map $\chi : \operatorname{ob}\mathcal{P}^{\{\mathfrak{m}\}}(R) \to \mathbb{Z}$ sending P to $\sum_i (-1)^i \operatorname{length}_R H_i(P)$ induces a homomorphism*

$$\chi : K_0^{\mathfrak{m}}(R) \to \mathbb{Z}.$$

If $(R, \mathfrak{m})$ is a regular local ring, this map is an isomorphism and $K_0^{\mathfrak{m}}(R)$ is the free abelian group of rank 1 generated by the class of the Koszul complex on a regular system of parameters.

Proof If P is acyclic then $\chi(P) = 0$. Using the long exact sequence in homology associated to a short exact sequence of chain complexes, we see that χ is also additive on short exact sequences of complexes. It thus induces the indicated homomorphism χ on $K_0^{\mathfrak{m}}(R)$.

Assume R is regular local and let $K \in \mathcal{P}^{\mathfrak{m}}(R)$ be the Koszul complex on a regular system of parameters. Then $H_0(K) = R/\mathfrak{m}$ and $H_i(K) = 0$ for all $i \neq 0$, and hence $\chi([K]) = 1$. In particular, χ is onto. Let E be any object of $\mathcal{P}^{\mathfrak{m}}(R)$. It remains to show $[E]$ is in the subgroup generated by $[K]$. We proceed by induction on $h = h(E) = \sum_i \operatorname{length}_R H_i(E)$. If $h = 0$, then E is acyclic and hence $[E] = [0] = 0$.

Assume $h > 0$. Let i be the largest integer such that $H_i(E) \neq 0$, and pick $\alpha \in H_i(E)$ such that $\alpha \neq 0$ and $\mathfrak{m}\alpha = 0$. Since $H_j(E) = 0$ for $j > i$, it follows that there is a chain map $g : \Sigma^i K \to E$ such that the induced map on H_i has the form $R/\mathfrak{m} \to H_i(E)$, sending 1 to α. Let $C = \operatorname{cone}(g)$. Then we have an exact sequence $0 \to E \to C \to \Sigma^{i+1} K \to 0$ so that $[E] = [C] - (-1)^{i+1}[K]$. By construction, $h(C) = h(E) - 1$ and thus by induction $[C] \in \mathbb{Z} \cdot [K]$. $\square$

Remark 4.5 A slight generalization of the proof just given shows, more generally, that if R is regular and semi-local with $\operatorname{mSpec}(R) = \{\mathfrak{m}_1, \ldots, \mathfrak{m}_n\}$, then the localization maps $R \to R_{\mathfrak{m}_i}$ induce an isomorphism

$$K_0^{\operatorname{mSpec}(R)}(R) \xrightarrow{\cong} \bigoplus_{i=1}^{n} K_0^{\mathfrak{m}_i}(R_{\mathfrak{m}_i}) \cong \mathbb{Z}^n.$$

Remark 4.6 An interesting question is whether the analogue of the previous Lemma holds for $\mathbb{Z}/2$-graded complexes.

Remark 4.7 When R is not regular, $K_0^\mathfrak{m}(R)$ is in general much bigger than $\mathbb{Z}$. It is typically not at all easy to compute its value, and it is typically not finitely generated. Here is one case in which it can be computed: Assume $(R, \mathfrak{m})$ is a local domain of dimension 1, and let F be its field of fractions. Then there is a homomorphism of abelian groups $F^\times \to K_0^\mathfrak{m}(R)$ defined by sending $\frac{a}{b} \in F^\times$ to $[R \xrightarrow{a} R] - [R \xrightarrow{b} R] \in K_0^\mathfrak{m}(R)$ for any $a, b \in R \setminus \{0\}$. (Note that $R \xrightarrow{c} R$ has finite length homology for any $0 \neq c \in R$ since $\dim(R) = 1$.) The kernel of this map clearly contains $R^\times$ and it can be shown that in fact it induces an isomorphism

$$\frac{F^\times}{R^\times} \cong K_0^\mathfrak{m}(R)$$

of abelian groups.

Note that if R is regular (i.e., a DVR) then the valuation mapping gives an isomorphism $F^\times / R^\times \cong \mathbb{Z}$, as the Lemma tells us.

4.2.3 Cup Product

Given a pair of closed subset Z and W of $\mathrm{Spec}(R)$, tensor product of complexes determines a bi-functor

$$\mathcal{P}^Z(R) \times \mathcal{P}^W(R) \to \mathcal{P}^{Z \cap W}(R)$$

given by $(P, P') \mapsto P \otimes_R P'$, that is bi-exact and preserves homotopy equivalences in each argument. It thus induces a bilinear pairing:

Definition 4.3 The *cup product* pairing is the map on Grothendieck groups

$$- \cup -: K_0^Z(R) \times K_0^W(R) \to K_0^{Z \cap W}(R)$$

given by $[P] \cup [P'] = [P \otimes_R P']$. (Here $P \otimes_R P'$ refers to the tensor product complex.)

This is a well-defined pairing, since $P \otimes_R -$ maps acyclic complexes to acyclic complexes, and short exact sequences to short exact sequences, and similarly for $- \otimes_R P'$. Moreover, for $P \in \mathcal{P}^Z(R)$, $P' \in \mathcal{P}^W(R)$, if $\mathfrak{p} \in \mathrm{Spec}(R) \setminus (Z \cap W)$ then either $P_\mathfrak{p}$ or $P'_\mathfrak{p}$ is exact, and hence so is $(P \otimes_R P')_\mathfrak{p} \cong P_\mathfrak{p} \otimes_{R_\mathfrak{p}} P'_\mathfrak{p}$. This shows that the target of $- \cup -$ is indeed $K_0^{Z \cap W}(R)$.

The following is easy to prove by checking on generators:

Lemma 4.4 *The cup product operation is commutative, associative, and unital, in the appropriate senses.*

Example 4.7 Suppose $(R, \mathfrak{m})$ is a local ring of dimension d and let $x_1, \ldots, x_d$ be a system of parameters. Let $\mathrm{Kos}_R(x_i) = (\cdots 0 \to R \xrightarrow{x_i} R \to 0 \to \cdots)$ be the Koszul complex on x_i, for each i. Then $[\mathrm{Kos}_R(x_i)] \in K_0^{V(x_i)}(R)$ for each i and we have an equation

$$[\mathrm{Kos}_R(x_1)] \cup \cdots \cup [\mathrm{Kos}_R(x_d)] = [\mathrm{Kos}_R(x_1, \ldots, x_d)] \in K_0^{\mathfrak{m}}(R),$$

since $V(x_1, \cdots, x_d) = \{\mathfrak{m}\}$.

Example 4.8 Given a finitely generated R-module M of finite projective dimension, a chosen bounded projective resolution P^M gives a class in $K_0^{\mathrm{supp}(M)}(R)$. For another such module N, we have $[P^N] \in K_0^{\mathrm{supp}(N)}(R)$ and

$$[P^M] \cup [P^N] = [M \otimes_R^{\mathbb{L}} N] \in K_0^{\mathrm{supp}(M) \cap \mathrm{supp}(N)}(R).$$

4.2.4 Intersection Multiplicity

Building on the previous example, suppose in addition that R is local and M and N are chosen such that $\mathrm{supp}(M) \cap \mathrm{supp}(N) \subseteq \{\mathfrak{m}\}$. (This is equivalent to $\mathrm{length}_R(M \otimes_R N) < \infty$.) Recall that there is a map

$$\chi : K_0^{\mathfrak{m}}(R) \to \mathbb{Z}$$

given by $\chi([P]) = \sum_i (-1)^i \mathrm{length}\, H_i(P)$. We get that

$$\chi([P_M] \cup [P_N]) = \sum_i (-1)^i \mathrm{length}_R(\mathrm{Tor}_i^R(M, N)) =: \chi(M, N).$$

This is *Serre's intersection multiplicity* formula.

The geometric intuition here is the following: Say $M = R/I$ and $N = R/J$ have finite projective dimension (e.g., say R is regular) and $\sqrt{I + J} = \mathfrak{m}$. Geometrically, $V(I)$ and $V(J)$ meet only at the single point $\{\mathfrak{m}\}$. The integer $\chi(R/I, R/J)$ gives the multiplicity of the intersection.

Example 4.9 Let k be a field and $R = k[[x, y]]$. Suppose $I = (f(x, y))$ and $J = (g(x, y))$ meet only at the maximal ideal. Then f, g have no common factors, and hence form a regular sequence, so that $\mathrm{Tor}_i^R(R/f, R/g) = 0$ for $i \neq 0$. We get

$$\chi(R/f, R/g) = \mathrm{length}_R R/(f, g) = \dim_k R/(f, g).$$

4.2.5 Functorality

Let $\phi: R \to S$ be a ring map and write $\phi^{\#}: \mathrm{Spec}(S) \to \mathrm{Spec}(R)$ be the induced map on spectra (i.e., $\phi^{\#}(\mathfrak{p}) = \phi^{-1}(\mathfrak{p})$). Suppose $Z \subseteq \mathrm{Spec}(R)$ and $W \subseteq \mathrm{Spec}(S)$ are closed subsets such that $(\phi^{\#})^{-1}(Z) \subseteq W$. (That is, assume that if $\mathfrak{q} \in \mathrm{Spec}(S) \setminus W$ then $\phi^{-1}(\mathfrak{q}) \in \mathrm{Spec}(R) \setminus Z$.) Then ϕ induces a homomorphism

$$\phi_* = \phi_*^{Z,W} : K_0^Z(R) \to K_0^W(S).$$

To see that the target is correct, suppose $\mathfrak{q} \in \mathrm{Spec}(S) \setminus W$ and $P \in \mathcal{P}^Z(R)$. By assumption $\mathfrak{p} = \phi^{-1}(\mathfrak{q}) \notin Z$ and thus

$$(P \otimes_R S)_{\mathfrak{q}} \cong P_{\mathfrak{p}} \otimes_{R_{\mathfrak{p}}} S_{\mathfrak{q}}$$

is acyclic.

Example 4.10 If $\phi = \mathrm{id}_R$ and $Z \subseteq W \subseteq \mathrm{Spec}(R)$, $K_0^Z(R) \to K_0^W(R)$ is induced by the inclusion $\mathcal{P}^Z(R) \subseteq \mathcal{P}^W(R)$. Beware that this map on Grothendieck groups is often not injective, since upon enlarging the support, one not only enlarges the number of generators but also the numbers of relations.

For example, if R is a local domain, $W = \mathrm{Spec}(R)$ and Z is any proper closed subset, then $K_0^Z(R) \to K_0^{\mathrm{Spec}(R)}(R)$ is the zero map: For recall that $K_0^{\mathrm{Spec}(R)}(R) \cong K_0(R) \cong \mathbb{Z}$, with the composition sending $[P]$ to $\sum_i (-1)^i \mathrm{rank}_R(P_i)$. If $P \in K_0^Z(R)$ and $Z \subseteq \mathrm{Spec}(R)$, then $P \otimes_R F$ is acyclic, where F is the field of fractions of R. It follows that $\sum_i (-1)^i \mathrm{rank}_R(P_i) = 0$.

Example 4.11 For any $f \in R$, we have the localization map $\phi: R \to R[1/f]$, which induces the map

$$K_0^Z(R) \to K_0^{Z \setminus V(f)}(R[1/f])$$

sending $[P]$ to $[P[1/f]]$.

4.2.6 A Right Exact Sequence

Theorem 4.2 (Gillet-Soulé) *For a regular ring R, closed subset Z of $\mathrm{Spec}(R)$, and element $f \in R$, the sequence*

$$K_0^{Z \cap V(f)}(R) \to K_0^Z(R) \to K_0^{Z \setminus V(f)}(R[1/f]) \to 0$$

is exact.

Proof *(Sketch of Proof)* The proof relies on the following fact: For any regular ring B and ideal J, we have an isomorphism $G_0(B/J) \xrightarrow{\cong} K_0^{V(J)}(B)$, where $G_0(B/J)$ denotes the Grothendieck group of all finitely generated B/J-modules and the map sends the class of such a module to the class of a projective resolution of it. The proof of this fact is not difficult, but I omit it.

Say $Z = V(I)$. Since R is regular, using the fact above, the sequence in the statement is isomorphic to the sequence

$$G_0(A/f) \to G_0(A) \to G_0(A[1/f]) \to 0,$$

where $A = R/I$, the first map is induced by restriction of scalars and the second by localization. The latter sequence is a portion of the well-known localization exact sequence in G-theory; I include a sketch of the proof, adapted from [19, ii.6.4].

The composition of $G_0(A/f) \to G_0(A) \to G_0(A[1/f])$ is the 0 map since $N[1/f] = 0$ for any A/f-module N. Given a finitely generated $A[1/f]$-module M, by choosing a presentation and clearing denominators, we can construct a finitely generated A-module N such that $N[1/f] \cong M$. It follows that $G_0(A) \to G_0(A[1/f])$ is onto.

Set $\Gamma = \operatorname{coker}(G_0(A/f) \to G_0(A))$. By what we have already shown, there is an induced surjection $\Gamma \to G_0(A[1/f])$, and we need to show it is an isomorphism. We do so by constructing an inverse map α. For each finitely generated $A[1/f]$-module M, as before choose a lift of it to a finitely generated A-module N such that there is an isomorphism $N[1/f] \cong M$ of $A[1/f]$-modules, and set $\alpha(M) \in \Gamma$ to be the image of the class $[N] \in G_0(A)$ under the canonical map $G_0(A) \twoheadrightarrow \Gamma$. Provided α induces a well-defined additive function that is independent of the choice of N, the induced map $\alpha \colon G_0(A[1/f]) \to \Gamma$ is a left inverse of $\Gamma \to G_0(A[1/f])$.

To see $\alpha(M)$ does not depend on the choice of N, suppose N' is another such lift. Then there exists an isomorphism $g \colon N[1/f] \xrightarrow{\cong} N'[1/f]$ of $A[1/f]$-modules. Multiplying though by a sufficiently high power of f, we may assume g lifts to a homomorphism $\tilde{g} \colon N \to N'$. Since $\ker(\tilde{g})$ and $\operatorname{coker}(\tilde{g})$ are annihilated by a power of f, both $[\ker(\tilde{g})]$ and $[\operatorname{coker}(\tilde{g})]$ lie in the image of $G_0(A/f) \to G_0(A)$. We see that $\alpha(M)$ is a well-defined function, independent of the choice of N.

Given a short exact sequence of finitely generated $A[1/f]$-modules $0 \to M' \to M \to M'' \to 0$, we may find finitely generated A-modules N, N' and maps $N' \to N, N \to N''$ that lift these. The composition of $N' \to N \to N''$ is not a priori 0, but by multiplying through by a power of f, this can be arranged. So, we have a complex $0 \to N' \to N \to N'' \to 0$, and its homology is annihilated by a power of f. As before, it follows that

$$\overline{[N]} = \overline{[N']} + \overline{[N'']}$$

holds in Γ.   □

4.2.7 Adams Operations

In order to better understand the groups $K_0^Z(R)$, we decompose them into so-called "weight pieces" that have certain desirable properties. These weight pieces are obtained by taking eigenspaces for operators that satisfy certain properties, and we call any such operator an "Adams operator". In the special case when $Z = \operatorname{Spec}(R)$, such a decomposition is given by Grothendieck's Theorem 4.1 using the classical Adams operator defined in Sect. 4.1.3. In the general case, we start by axiomatizing the needed properties:

Definition 4.4 (Gillet-Soulé) Let C be a collection of commutative noetherian rings and let k be a positive integer. An *Adams operation of degree k defined on* C is a collection of functions

$$\psi^k = \psi_{R,Z}^k : K_0^Z(R) \to K_0^Z(R)$$

for all $R \in C$ and all closed subsets $Z \subseteq \operatorname{Spec}(R)$ such that four axioms hold:

1. (Additivity) Each $\psi_{R,Z}^k$ is an endomorphism of abelian groups.
2. (Multiplicativity) For any $R \in C$ and closed subsets Z, W of $\operatorname{Spec}(R)$, the diagram

$$
\begin{array}{ccc}
K_0^Z(R) \times K_0^W(R) & \xrightarrow{\ \cup\ } & K_0^{Z \cap W}(R) \\
\Big\downarrow{\psi_{R,Z}^k \times \psi_{R,W}^k} & & \Big\downarrow{\psi_{R,Z \cap W}^k} \\
K_0^Z(R) \times K_0^W(R) & \xrightarrow{\ \cup\ } & K_0^{Z \cap W}(R)
\end{array}
$$

   commutes.
3. (Naturality) Given a ring homomorphism $\phi \colon R \to S$, with $R, S \in C$, and closed subsets $Z \subseteq \operatorname{Spec}(R)$, and $W \subseteq \operatorname{Spec}(S)$ such that $(\phi^{\#})^{-1}(Z) \subseteq W$, the diagram

$$
\begin{array}{ccc}
K_0^Z(R) & \xrightarrow{\ \phi_*^{Z,W}\ } & K_0^W(S) \\
\Big\downarrow{\psi_{R,Z}^k} & & \Big\downarrow{\psi_{S,W}^k} \\
K_0^Z(R) & \xrightarrow{\ \phi_*^{Z,W}\ } & K_0^W(S)
\end{array}
$$

   commutes.
4. (Normalization) For all $R \in C$ and $a \in R$

$$\psi_{R,V(a)}^k([\operatorname{Kos}_R(a)]) = k \cdot [\operatorname{Kos}_R(a)] \in K_0^{V(a)}(R),$$

where $\mathrm{Kos}_R(a) = \left(\cdots \to 0 \to R \xrightarrow{a} R \to 0 \to \cdots \right)$, the Koszul complex on a.

Example 4.12 Given an Adams operation on C of degree k, suppose $a_1, \ldots, a_c \in R$ and $R \in C$, and let

$$\mathrm{Kos}_R(a_1, \ldots, a_c) = \bigotimes_i \mathrm{Kos}_R(a_i).$$

The multiplicative and normalization axioms give

$$\psi^k([\mathrm{Kos}_R(a_1, \ldots, a_c)]) = k^c[\mathrm{Kos}_R(a_1, \ldots, a_c)] \in K_0^{V(a_1, \ldots, a_c)}(R).$$

4.2.8 Frobenius

One example of an Adams operation comes from the Frobenius:

Let p be a prime and C_p the collection of all commutative noetherian rings of characteristic p. For each $R \in C_p$, let $F: R \to R$ denote the Frobenius endomorphism. Since $F^\sharp: \mathrm{Spec}(R) \to \mathrm{Spec}(R)$ is the identity map of topological spaces, for each closed subset Z we have an induced map

$$F_*: K_0^Z(R) \to K_0^Z(R)$$

that sends $[P]$ to $[P \otimes_R {}^F R]$.

Proposition 4.2 F_* *is an Adams operation of degree p defined on C_p.*

Proof Axioms 1 and 2 hold since F_* is the homomorphism induced by extension of scalars. Axiom 3 holds by the naturality of Frobenius. For Axiom 4, we have

$$\mathrm{Kos}_R(a) \otimes_R {}^F R \cong \mathrm{Kos}_R(a^p)$$

and so it suffices to prove $[\mathrm{Kos}_R(a^p)] = p[\mathrm{Kos}_R(a)] \in K_0^{V(a)}(R)$ — this actually holds for any ring R, element $a \in R$ and integer p:

Let $\alpha: \mathrm{Kos}_R(a^{p-1}) \to \mathrm{Kos}_R(a^p)$ be the chain map given as the identity in degree 1 and multiplication by a in degree 0. Then $\mathrm{cone}(\alpha) \sim \mathrm{Kos}_R(a)$ and hence

$$[\mathrm{Kos}_R(a^p)] = [\mathrm{Kos}_R(a^{p-1})] + [\mathrm{Kos}_R(a)].$$

The result follows by induction on p. $\square$

4.3 Existence of Adams Operations, Weight Decompositions

4.3.1 The Adams Operations of Gillet-Soulé

Theorem 4.3 (Gillet-Soulé) *For each $k \geq 1$, there exists an Adams operation ψ_{GS}^k of degree k defined on the collection of all commutative noetherian rings. Moreover,*

$$\psi_{GS}^k \circ \psi_{GS}^j = \psi_{GS}^{jk}.$$

for all $j, k \geq 1$. In particular, any two such operations commute.

Some comments:

- They first establish λ operations on $K_0^Z(R)$ and define ψ_{GS}^k via the same formula used for classical K_0.
- These λ operations are defined by replacing complexes with simplicial modules (Dold-Kan correspondence) and taking exterior powers. This idea originated with Dold and Puppe.
- A difficult argument is needed to verify the axioms of a λ ring.

I will not prove the theorem of Gillet-Soulé.

4.3.2 Cyclic Adams Operations

I present another way to create Adams operations, called "cyclic Adams operations". The details may be found in [4].

An overview:

- The construction requires fixing a prime p and it yields just a single Adams operation, of degree p, which is defined only on the category of $\mathbb{Z}[1/p, \zeta_p]$-algebras, where ζ_p is a primitive p-th root of unity.
- These limitations are minor, at least if one is working locally: Starting with an arbitrary local ring R, one may pick any prime $p \notin \mathfrak{m}$ and pass to a finite étale extension $R \subseteq R'$ with $\zeta_p \in R'$. For most applications, no important information is lost by this process and so these operations are usually good enough.
- One advantage of the cyclic Adams operations, as compared with those defined by Gillet and Soulé, is that ψ_{cy}^p can be defined in other contexts too; e.g., on the Grothendieck group of matrix factorizations.

Given a bounded complex P of finitely generated, projective R-modules and an integer $n \geq 0$, define

$$T^n(P) := \overbrace{P \otimes_R \cdots \otimes_R P}^{n}$$

The symmetric group Σ_n acts on $T^n(P)$ by permuting the tensor factors and adhering to the Koszul sign convention:

$$\tau \cdot (x_1 \otimes \cdots \otimes x_n) = \pm x_{\tau(1)} \otimes \cdots \otimes x_{\tau(n)}$$

where the x_i's are elements belonging to the various graded components of the complex P. The sign is uniquely determined by the following rule: When τ is the adjacent transposition $\tau = (i \; i+1)$ for some $1 \le i \le n-1$, we have

$$\tau \cdot (x_1 \otimes \cdots \otimes x_n) = (-1)^{|x_i||x_{i+1}|} x_1 \otimes \cdots \otimes x_{i-1} \otimes x_{i+1} \otimes x_i \otimes x_{i+2} \otimes \cdots \otimes x_n,$$

where if $x \in P_m$ we set $|x| = m$. For a general τ, the sign is $(-1)^e$ where $e = \sum_{i<j, \tau^{-1}(i)>\tau^{-1}(j)} |x_i||x_j|$.

This sign rule gives that the action of Σ_n commutes with the differential on $T^n(P)$; that is, we may regard $T^n(P)$ as a complex of left modules over the group ring $R[\Sigma_n]$.

Let $C_n = \langle \sigma \rangle$ be the cyclic subgroup of Σ_n of order n generated by the cyclic permutation $\sigma = (1\,2\,\cdots\,n)$. Then C_n acts on $T^n(P)$ by

$$\sigma \cdot (x_1 \otimes \cdots \otimes x_n) = (-1)^{|x_1|(|x_2|+\cdots+|x_n|)} x_2 \otimes \cdots \otimes x_n \otimes x_1$$

Now assume $n = p$, for a prime p, and that R contains $\frac{1}{p}$ and a primitive p-th root of unity ζ_p. These assumptions give that the group ring $R[C_p] \cong R[x]/(x^p - 1)$ decomposes as a product of copies of R, indexed by the primitive p-th roots of unity, and hence that there is an internal direct sum decomposition

$$T^p(P) = \bigoplus_{j=0}^{p-1} T^p(P)^{(\zeta_p^j)}$$

of complexes of R-modules, where we set

$$T^p(P)^{(\zeta_p^j)} = \ker(T^p(P) \xrightarrow{\sigma - \zeta_p^j} T^p(P)).$$

I will be primarily interested in the case $p = 2$. Note that $\zeta_2 = -1$ and for $P \in \mathcal{P}^{\mathbb{Z}}(R)$ the group $\Sigma_2 = C_2 = \langle \sigma \rangle$ acts on $T^2(P)$ by $\sigma \cdot (x \otimes y) = (-1)^{|x||y|} y \otimes x$. So,

$$T^2(P)^{(1)} = S^2(P) := \{\alpha \in P \otimes_R P \mid \sigma \cdot \alpha = \alpha\}$$

and

$$T^2(P)^{(-1)} = \Lambda^2(P) := \{\alpha \in P \otimes_R P \mid \sigma \cdot \alpha = -\alpha\}.$$

Since we assume $\frac{1}{2} \in R$, we have an internal direct sum decomposition

$$T^2(P) = S^2(P) \oplus \Lambda^2(P)$$

of chain complexes. In particular, we have

$$\mathrm{supp}(S^2(P)) \subseteq \mathrm{supp}(P) \text{ and } \mathrm{supp}(\Lambda^2(P)) \subseteq \mathrm{supp}(P). \tag{4.8}$$

For example, if P is a projective module viewed as a complex with trivial differential concentrated in even degree d, then we may identify $S^2(P)$ and $\Lambda^2(P)$ with the classical second symmetric and exterior powers of P, viewed as a complex in degree $2d$. In the same situation but with d odd, the roles are flipped. In general, the graded module underlying $S^2(P)$ for an arbitrary P is the tensor product of the classical second symmetric power of the even part of P with the classical second exterior power of the odd part, and vice versa for $\Lambda^2(P)$.

Proposition 4.3 (Brown-Miller-Thompson-Walker) *Assume p is a prime and that $\frac{1}{p}, \zeta_p \in R$. For any Z, there is a well-defined endomorphism of abelian groups*

$$\psi_{cy}^p \colon K_0^Z(R) \to K_0^Z(R)$$

given on generators by the formula

$$\psi_{cy}^p([P]) = [T^p(P)^{(1)}] - [T^p(P)^{(\zeta_p)}].$$

In particular, for $p = 2$, we have the operator

$$\psi_{cy}^2([P]) = [S^2(P)] - [\Lambda^2(P)].$$

Definition 4.5 The function ψ_{cy}^p described in the Proposition is called the *p-th cyclic Adams operation*.

Proof *(When $p = 2$)* For $P \in \mathcal{P}^Z(R)$ set $\psi_{cy}^2(P) = [S^2(P)] - [\Lambda^2(P)] \in K_0^Z(R)$. Note that $\psi_{cy}^2(P)$ does indeed belong to $K_0^Z(R)$ by (4.8). We must show this function respects the two defining relations of $K_0^Z(R)$.

Given a quasi-isomorphism $P \xrightarrow{\sim} P'$, we have a quasi-isomorphism $P \otimes_R P \xrightarrow{\sim} P' \otimes_R P'$. Since $\frac{1}{2} \in R$, this map decomposes as a direct sum of maps of the form $S^2(P) \to S^2(P')$ and $\Lambda^2(P) \to \Lambda^2(P')$, and each of these must also be quasi-isomorphisms. This proves $\psi_{cy}^2(P) = \psi_{cy}^2(P')$.

Suppose $0 \to P' \xrightarrow{i} P \to P'' \to 0$ is a short exact sequence in $\mathcal{P}^Z(R)$. We consider the filtration of complexes

$$0 = F_0 \subseteq F_1 \subseteq F_2 \subseteq F_3 = P \otimes_R P$$

where $F_1 = P' \otimes_R P'$ and $F_2 = P' \otimes_R P + P \otimes_R P'$. (Here, we interpret i as an inclusion of a subcomplex P' into P.) This filtration respects the action of C_2, and we have a C_2-equivariant isomorphism of complexes

$$F_2/F_1 \cong P' \otimes_R P'' \oplus P'' \otimes_R P'$$

where the C_2 action on the right is given by

$$\sigma(x' \otimes x'' + y'' \otimes y') = (-1)^{|y'||y''|} y' \otimes y'' + (-1)^{|x'||x''|} x'' \otimes x'.$$

Taking invariants yields the isomorphisms

$$(F_2/F_1)^{(1)} \cong P' \otimes_R P''$$
$$(F_2/F_1)^{(-1)} \cong P' \otimes_R P''.$$

Likewise, we have C_2-equivariant isomorphisms $F_3/F_2 \cong P'' \otimes_R P''$ and $F_1/F_0 \cong P' \otimes_R P'$ and hence isomorphisms

$$(F_1/F_0)^{(1)} \cong S^2(P')$$
$$(F_1/F_0)^{(-1)} \cong \Lambda^2(P')$$
$$(F_3/F_2)^{(1)} \cong S^2(P'')$$
$$(F_3/F_2)^{(-1)} \cong \Lambda^2(P'').$$

These isomorphisms give the equations

$$[S^2(P)] = [S^2(P')] + [P' \otimes_R P''] + [S^2(P'')]$$

and

$$[\Lambda^2(P)] = [\Lambda^2(P')] + [P' \otimes_R P''] + [\Lambda^2(P'')]$$

in $K_0^Z(R)$. Taking their difference yields $\psi_{cy}^2(P) = \psi_{cy}^2(P') + \psi_{cy}^2(P'')$.   □

Theorem 4.4 (Brown-Miller-Thompson-Walker) ψ_{cy}^p *satisfies the four Gillet-Soulé axioms on the category* C_p.

Proof *(Sketch When $p = 2$)* The first axiom, additivity, is given by Proposition 4.3.
   I omit a proof of the second axiom, multiplicativity, although it is not too hard.
   The third axiom, naturality, is a consequence of the naturality of the functors S^2 and Λ^2.
   Let $a \in R$ and set $K = \mathrm{Kos}_R(a)$. Then $T^2(K) = \mathrm{Kos}_R(a, a) \cong \mathrm{Kos}_R(a, 0)$. Explicitly, if K has R-basis α, β with $|\alpha| = 1$, $|\beta| = 0$ and $d(\alpha) = a\beta$, then $\mathrm{Kos}_R(a, 0)$ has basis $x = \alpha \otimes \alpha$, $y = \alpha \otimes \beta + \beta \otimes \alpha$, $z = \beta \otimes \alpha - \alpha \otimes \beta$,

and $w = \beta \otimes \beta$, of degrees 2, 1, 1, and 0, respectively. Relative to this basis, we compute the action of the differential and the transposition operators: $d(x) = az$, $d(y) = 2aw$, $d(z) = 0$, and $d(w) = 0$, and $\sigma \cdot x = -x$, $\sigma \cdot y = y$, $\sigma \cdot z = -z$, and $\sigma \cdot w = w$. It follows that

$$S^2(K) \cong \mathrm{Kos}_R(2a)$$

and

$$\Lambda^2(K) \cong \Sigma \, \mathrm{Kos}_R(-a).$$

Hence

$$\psi_{cy}^2(K) = [\mathrm{Kos}_R(2a)] + [\mathrm{Kos}_R(-a)] = [\mathrm{Kos}_R(a)] + [\mathrm{Kos}_R(a)] = 2[K].$$

This establishes the final axiom. □

It is not known (at least to me) if the cyclic Adams operations coincide with those defined by Gillet and Soulé in all situations in which the former are defined. But the following is proven in [4]:

Proposition 4.4 (Brown-Miller-Thompson-Walker) *Let R be a noetherian ring and p a prime. If $p!$ is invertible in R, then ψ_{cy}^p and ψ_{GS}^p coincide as operators on $K_0^Z(R)$ for all Z. In particular, $\psi_{cy}^2 = \psi_{GS}^2$ for local rings $(R, \mathfrak{m})$ with $\mathrm{char}(R/\mathfrak{m}) \neq 2$.*

4.3.3 The Weight Decomposition

As mentioned, the essential feature of Adams operations is that they decompose rationalized Grothendieck groups of regular rings into "weight pieces". The precise statement is:

Theorem 4.5 (Gillet-Soulé) *Assume ψ^k is an Adams operation of degree k, for any $k \geq 2$, defined on some collection of commutative rings C that is closed under localization. If $R \in C$ is regular and $Z = V(I) \subseteq R$ is any closed subset, then there exists an internal direct sum decomposition*

$$K_0^Z(R) \otimes_{\mathbb{Z}} \mathbb{Q} = \bigoplus_{j=\mathrm{height}(I)}^{\dim(R)} K_0^Z(R)_{\mathbb{Q}}^{(j)}$$

where we define

$$K_0^Z(R)_{\mathbb{Q}}^{(j)} := \ker(K_0^Z(R) \otimes_{\mathbb{Z}} \mathbb{Q} \xrightarrow{\psi^k \otimes \mathbb{Q} - k^j} K_0^Z(R) \otimes_{\mathbb{Z}} \mathbb{Q}).$$

In other words, the theorem says that the operator $\psi^k \otimes_{\mathbb{Z}} \mathbb{Q}$ is diagonalizable with eigenvalues contained in the set $\{k^j \mid \text{height}(I) \leq j \leq \dim(R)\}$. In yet other words, $K_0^Z(R)_{\mathbb{Q}}$ is annihilated by $\prod_{\text{height}(I) \leq j \leq \dim(R)} (\psi^k - k^j)$.

Definition 4.6 The subspace $K_0^Z(R)_{\mathbb{Q}}^{(j)}$ is the j-th *weight space* for the operator ψ^k. To indicate the (possible) dependence on the choice of operator, we sometimes write this $K_0^Z(R)_{\mathbb{Q}}^{(j)_{\psi^k}}$.

Remark 4.8 Before proving this Theorem, I make an observation that is simple but important in applications: The weight decomposition is multiplicative. That is, for all R, Z, W, i, j, for a given Adams operation the cup product pairing induces a pairing on the associate weight pieces of the form

$$K_0^Z(R)_{\mathbb{Q}}^{(i)} \times K_0^W(R)_{\mathbb{Q}}^{(j)} \to K_0^{Z \cap W}(R)_{\mathbb{Q}}^{(i+j)}.$$

This is immediate from Axiom 2 and the definitions.

Proof *(Sketch of Proof)* Let us regard $K_0^{V(I)}(R)_{\mathbb{Q}}$ as a $\mathbb{Q}[t]$-module with t acting as ψ^k. We need to prove it is annihilated by the polynomial $\prod_{\text{height}(I) \leq j \leq \dim(R)} (t - k^j)$.

If this is false for some regular noetherian ring R, then we may find an ideal I that is maximal among those ideals for which it fails to hold. Let $\mathfrak{p}_1, \ldots, \mathfrak{p}_n$ be the minimal primes that contain I and are of the same height as I. Set $S = R \setminus (\mathfrak{p}_1 \cup \cdots \cup \mathfrak{p}_n)$. By taking colimits over all $g \in S$ in Theorem 4.2, we obtain the right exact sequence

$$\bigoplus_{g \in S} K_0^{V(I,g)}(R)_{\mathbb{Q}} \to K_0^{V(I)}(R)_{\mathbb{Q}} \to K_0^{\text{Spec}(S^{-1}R) \cap V(I)}(S^{-1}R)_{\mathbb{Q}} \to 0 \qquad (4.9)$$

of $\mathbb{Q}[t]$-modules (i.e., the maps commute with the action of ψ^k by Axiom 3).

By choice of I, the summand indexed by g in the left-most term is annihilated by $\prod_{\text{height}(I,g) \leq j \leq \dim(R)} (t^k - k^j)$, and so, since $\text{height}(I, g) > \text{height}(I)$ for each $g \in S$, the left-most term is annihilated by $\prod_{\text{height}(I)+1 \leq j \leq \dim(R)} (t^k - k^j)$.

By choice of S, $S^{-1}R$ is semi-local, $\text{Spec}(S^{-1}R) \cap V(I) = \text{mSpec}(S^{-1}R)$, and each prime in $\text{mSpec}(S^{-1}R)$ has height equal to $\text{height}(I)$. By Remark 4.5 we have an isomorphism

$$K_0^{\text{mSpec}(S^{-1}R)}(S^{-1}R)_{\mathbb{Q}} \cong \bigoplus_{i=1}^n K_0^{\mathfrak{p}_i R_{\mathfrak{p}_i}}(R_{\mathfrak{p}_i})$$

and it is compatible with the action of ψ^k by Axiom 3. Since $R_{\mathfrak{p}_i}$ is regular local of dimension $\text{height}(I)$ for each i, by Lemma 4.3 $K_0^{\mathfrak{p}_i R_{\mathfrak{p}_i}}(R_{\mathfrak{p}_i})$ is generated by the class of a Koszul complex on $\text{height}(I)$ elements and thus, by Example 4.12, it is annihilated by $t^k - k^{\text{height}(I)}$.

We have proven that the right-most term of (4.9) is annihilated by $t^k - k^{\text{height}(I)}$ and its left-most term is annihilated by $\prod_{\text{height}(I)+1 \leq j \leq \dim(R)}(t^k - k^j)$. Since these polynomials are relatively prime elements of $\mathbb{Q}[t]$, it follows that the middle term $K_0^{V(I)}(R)_{\mathbb{Q}}$ is annihilated by their product, namely $\prod_{\text{height}(I) \leq j \leq \dim(R)}(t - k^j)$. We have reached a contradiction. $\qquad\square$

Remark 4.9 This proof does not show that $K_0^{\mathcal{Z}}(R)_{\mathbb{Q}}$ admits such a weight decomposition for an arbitrary noetherian ring R. The most important things that fails in the proof is that the groups $K_0^{\mathfrak{p}_i R_{\mathfrak{p}_i}}(R_{\mathfrak{p}_i})$ are not necessarily of "pure weight"—i.e., ψ^k need not act as multiplication by k^h on these groups.

Remark 4.10 Assume ψ^k, ϕ^j are degree k, j Adams operation with $k, j \geq 2$. If ψ^k and ϕ^j commute, then

$$K_0^{\mathcal{Z}}(R)_{\mathbb{Q}}^{(j)_{\psi^k}} = K_0^{\mathcal{Z}}(R)_{\mathbb{Q}}^{(j)_{\phi^j}}.$$

In particular if $j = k$ and ψ^k and ϕ^k commute, then the operators themselves coincide at least rationally: $\psi_{\mathbb{Q}}^k = \phi_{\mathbb{Q}}^k$.

It is not known (at least to me) if the assumption that ψ^k, ϕ^j commute is needed for these facts to hold. That is, it is unknown if, for a fixed $k \geq 2$, the Gillet-Soulé axioms uniquely specify the rational degree k Adams operator for a regular ring.

However, if p is a prime and C is the category of commutative noetherian rings of characteristic p, then for any Adams operation ψ^k of degree k defined on C, we have

$$K_0^{\mathcal{Z}}(R)_{\mathbb{Q}}^{(j)_{\psi^k}} = K_0^{\mathcal{Z}}(R)_{\mathbb{Q}}^{(j)_F}$$

where F is the degree p Adams operation induced by Frobenius. In particular, $\psi_{\mathbb{Q}}^p = F_{\mathbb{Q}}$ on C.

4.4 Applications to Some Conjectures in Commutative Algebra

In this section I derive theorems in local algebra using the properties of Adams operations established above.

4.4.1 Serre Vanishing Conjecture

In this subsection I present the proof of the Serre Vanishing Conjecture (for regular rings) due to Gillet and Soulé. The proof I give is actually a slight modification of

their original proof, in that it uses the cyclic Adams operations instead of the ones
they used.

Let us recall the definition of the intersection pairing:

Definition 4.7 Let $(R, \mathfrak{m})$ be a local ring, and let M and N be finitely generated R-modules such that $\text{supp}(M) \cap \text{supp}(N) \subseteq \{\mathfrak{m}\}$ (or, equivalently, $\text{length}_R(M \otimes_R N) < \infty$). Assume also that either $\text{pd}_R(M) < \infty$ or $\text{pd}_R(N) < \infty$. Then the *intersection multiplicity* of M and N is the integer

$$\chi(M, N) = \sum_j (-1)^j \text{ length Tor}_j^R(M, N).$$

Conjecture 4.1 (Serre) Suppose R is a regular local ring and M and N are finitely generated R-modules such that $\text{supp}(M) \cap \text{supp}(N) \subseteq \{\mathfrak{m}\}$. If $\dim(M) + \dim(N) < \dim(R)$, then $\chi(M, N) = 0$.

Heuristically, the conjecture predicts that if the supports of M and N "ought not" meet, then their intersection multiplicity is 0.

Example 4.13 Two curves in three space meeting at the origin should have intersection multiplicity 0. That is, for the ring $R = k[[x, y, z]]$, given prime ideals $\mathfrak{p}, \mathfrak{q}$ such that $\dim(R/\mathfrak{p}) = 1$, $\dim(R/\mathfrak{q}) = 1$ and $\mathfrak{p} + \mathfrak{q}$ is $\mathfrak{m}$-primary, then we expect $\chi(R/\mathfrak{p}, R/\mathfrak{q}) = 0$. Let's check this holds for a pair of "axes": Let $\mathfrak{p} = (x, y), \mathfrak{q} = (y, z)$. We may resolve each by Koszul complexes and so their derived tensor product is $\text{Kos}_R(x, y, y, z) \cong \text{Kos}_R(x, y, z, 0)$, which is quasi-isomorphic to the complex $\cdots \to 0 \to k \xrightarrow{0} k \to 0 \to \cdots$, and hence $\chi = 0$.

Remark 4.11 The Vanishing Conjecture was proven by Serre himself when R contains a field.

The conjecture admits an evident generalization to the case when R is local, but not necessarily regular, provided either $\text{pd}_R(M) < \infty$ *or* $\text{pd}_R(N) < \infty$. It is known to be false at that level of generality; see the famous example constructed by Dutta, Hochster, and McLaughlin [7]. It remains an open conjecture for an arbitrary local ring R if it is assumed that *both* M and N have finite projective dimension. This has been proven to hold when R is a complete intersection ring by Roberts [14] and, independently, by Gillet and Soulé [9]. Roberts [op. cit.] has also proven it for isolated singularities.

Proof *(The Gillet-Soulé Proof of SVC for All Regular Local Rings)* Let R be a regular local ring, M, N finitely generated R-modules such that $\text{length}(M \otimes_R N) < \infty$ and $\dim(M) + \dim(N) < d = \dim(R)$. We show $\chi(M, N) = 0$ by following Gillet and Soulé's proof closely, except that we will use the cyclic Adams operations. This requires the addition of one preliminary step:

Pick a prime integer p such that $p \neq \text{char}(R/\mathfrak{m})$, so that $\frac{1}{p} \in R$. There exists a finite étale extension $R \subseteq R'$ of regular local rings such that R' contains a primitive

p-th root of unity ζ_p. For such an extension we have

$$\chi_{R'}(M \otimes_R R', N \otimes_R R') = \dim_{R/\mathfrak{m}}(R'/\mathfrak{m}') \cdot \chi_R(M, N).$$

We may thus assume without loss of generality that R contains ζ_p, so that the cyclic Adams operator ψ_{cy}^p is defined.

Choose projective resolutions P_M and P_N of M and N. Then

$$[P_M] \in K_0^{\text{supp}(M)}(R) \otimes_{\mathbb{Z}} \mathbb{Q} = \bigoplus_{j=d-\dim(M)}^{d} K_0^{\text{supp}(M)}(R)_{\mathbb{Q}}^{(j)}$$

by Theorem 4.5, and similarly for $[P_N]$. Since the cup product respects the weight decomposition (see Remark 4.8), we have

$$[P_M] \cup [P_N] = [P_M \otimes_R P_N] \in \bigoplus_{i \geq d-\dim(M), j \geq d-\dim(N)} K_0^{\mathfrak{m}}(R)_{\mathbb{Q}}^{(i+j)}.$$

The assumption $\dim(M) + \dim(N) < d$ implies $i + j > d$ for all such pairs, but we also know that $K_0^{\mathfrak{m}}(R) \otimes_{\mathbb{Z}} \mathbb{Q} = K_0^{\mathfrak{m}}(R)_{\mathbb{Q}}^{(d)}$ and hence

$$[P_M] \cup [P_N] = 0 \in K_0^{\mathfrak{m}}(R)_{\mathbb{Q}}.$$

Recall $K_0^{\mathfrak{m}}(R)_{\mathbb{Q}} \cong \mathbb{Q}$ and under this isomorphism $[P_M] \cup [P_N]$ corresponds to $\chi(M, N) \in \mathbb{Z} \subseteq \mathbb{Q}$. We have thus shown that $\chi(M, N) = 0$. □

4.4.2 The Total Rank Conjecture

Let $(R, \mathfrak{m})$ be a local ring with residue field $k = R/\mathfrak{m}$, and assume M is an R-module of finite length such that $M \neq 0$ and $\text{pd}_R(M) < \infty$. Let $\beta_i^R(M)$ denote the i-th Betti number of M, defined as $\beta_i^R(M) = \dim_k \text{Tor}_i^R(M, k)$ or, equivalently, as the rank of the i-th free module in the minimal free resolution of M over R. A famous conjecture attributed to Buchsbaum-Eisenbud [6] and Horrocks [11, Problem 24] predicts that $\beta_i^R(M) \geq \binom{d}{i}$ where $d = \dim(R)$. It remains open even for regular local rings and for polynomial rings in the graded setting.

Since $\sum_i \binom{d}{i} = 2^d$, an evident consequence of the BEH conjecture is:

Conjecture 4.2 (Avramov's Total Rank Conjecture) For a local ring R and a non-zero R-module M of finite length and finite projective dimension, we have

$$\sum_i \beta_i^R(M) \geq 2^{\dim(R)}.$$

The following theorem is from [18]:

Theorem 4.6 (Walker) *Let* $(R, \mathfrak{m})$ *be a regular local ring such that* char$(R/\mathfrak{m}) \neq$ 2. *If* P *belongs to* $\mathcal{P}^{\mathfrak{m}}(R)$ *and* P *is not exact, then*

$$\text{rank}_R(P) \geq 2^{\dim(R)} \frac{|\chi(P)|}{h(P)},$$

where $\text{rank}_R(P) = \sum_i \text{rank}_R(P_i)$, $\chi(P) = \sum_i (-1)^i \text{length}_R H_i(P)$, *and* $h(P) = \sum_i \text{length}_R H_i(P)$.

In particular, the Total Rank Conjecture holds for R.

Proof We easily deduce the second assertion from the first: Given a non-zero R-module M of finite length, let P be its minimal free resolution. Then we have $\text{rank}_R(P) = \sum_i \beta_i^R(M)$ and $\chi(P) = h(P) = \text{length}_R(M)$, and thus the first assertion gives $\sum_i \beta_i^R(M) \geq 2^{\dim(R)} \frac{|\chi(P)|}{h(P)} = 2^{\dim(R)}$.

Since char$(R/\mathfrak{m}) \neq 2$, R contains $\frac{1}{2}$ (and, obviously, $\zeta_2 = -1$), and thus the cyclic Adams operation ψ_{cy}^2 is defined for R and all of its localizations. Since R is regular, ψ_{cy}^2 acts as $2^{\dim(R)}$ on $K_0^{\mathfrak{m}}(R)$ and thus

$$\chi(\psi_{cy}^2(P)) = 2^{\dim(R)} \cdot \chi(P).$$

Recall that $\psi_{cy}^2(P) = [S^2(P)] - [\Lambda^2(P)]$ so that

$$\chi(\psi_{cy}^2(P)) = \chi(S^2(P)) - \chi(\Lambda^2(P))$$

$$\leq \sum_{i \text{ even}} \text{length}_R H_i(S^2 P) + \sum_{i \text{ odd}} \text{length}_R H_i(\Lambda^2 P)$$

$$\leq h(S^2(P)) + h(\Lambda^2(P)).$$

Since $P \otimes_R P = S^2(P) \oplus \Lambda^2(P)$, we have $h(P \otimes_R P) = h(S^2(P)) + h(\Lambda^2(P))$ and hence

$$\chi(\psi_{cy}^2(P)) \leq h(P \otimes_R P).$$

I claim that

$$h(P \otimes_R P) \leq \text{rank}(P)h(P). \tag{4.10}$$

To see this, consider the convergent spectral sequence

$$E_{i,j}^2 = H_i(P \otimes_R H_j(P)) \Longrightarrow H_{i+j}(P \otimes_R P)$$

that arises from regarding $P \otimes_R P$ as the totalization of a bicomplex. Since $\text{length}_R E_{i,j}^\infty \leq \text{length}_R H_i(P \otimes_R H_j(P))$ for all i, j and $\text{length}_R H_n(P \otimes_R P) =$

$\sum_{i+j=n}$ length$_R$ $E^\infty_{i,j}$, we deduce

$$\text{length}_R\, H_n(P \otimes_R P) \leq \sum_{i,j;i+j=n} \text{length}_R\, H_i(P \otimes_R H_j(P))$$

for each n. Now, $H_i(P \otimes_R H_j(P))$ is a subquotient of the finite length module $P_i \otimes_R H_j(P)$, and hence

$$\text{length}_R\, H_i(P \otimes_R H_j(P)) \leq \text{rank}_R(P_i)\, \text{length}(H_j(P)). \tag{4.11}$$

We conclude

$$h(P \otimes_R P) \leq \sum_n \sum_{i+j=n} \text{rank}_R(P_i)\, \text{length}(H_j(P))$$

$$= \sum_{i,j} \text{rank}_R(P_i)\, \text{length}_R(H_j(P))$$

$$= \text{rank}(P)h(P).$$

To conclude the proof, we just put the above inequalities together to get

$$\text{rank}(P)h(P) \geq 2^{\dim(R)}\chi(P),$$

and since $h(P) > 0$ we conclude

$$\text{rank}_R(P) \geq 2^{\dim(R)}\frac{\chi(P)}{h(P)}.$$

Applying this inequality to ΣP in place P, and using that $\text{rank}_R(\Sigma P) = \text{rank}_R(P)$, $h(\Sigma P) = h(P)$, and $\chi(\Sigma P) = -\chi(P)$, we also get $\text{rank}_R(P) \geq -2^{\dim(R)}\frac{\chi(P)}{h(P)}$. $\qquad\square$

Remark 4.12 The inequality (4.10) can also be proven without using spectral sequences. More generally, if M is any bounded complex of R-modules having finite length homology, I claim $h(P \otimes_R M) \leq \text{rank}(P)h(M)$. Both sides are unaffected by replacing M with any complex that is quasi-isomorphic to it. In particular, it holds when M is acyclic. If M is not acyclic, proceed by induction on $w = w(M) := \max\{n \mid H_n(M) \neq 0\} - \min\{n \mid H_n(M) \neq 0\}$. When $w = 0$, M is quasi-isomorphic to a module viewed as a complex concentrated in one degree. In this case the inequality holds for the same reason (4.11) holds. For $w > 0$, by using a "soft truncation", we may find a short exact sequence

$$0 \to M' \to M \to M'' \to 0 \tag{4.12}$$

of non-acyclic, bounded complexes with finite length homology such that the map induced map on homology $H(M) \rightarrow H(M'')$ is surjective, $w(M') < w(M)$, and $w(M'') < w(M)$. These conditions imply that $h(M) = h(M') + h(M'')$. Moreover, the long exact sequence in homology associated to the short exact sequence obtained by applying $P \otimes_R -$ to (4.12) yields the inequality

$$h(P \otimes_R M) \leq h(P \otimes_R M') + h(P \otimes_R M'').$$

By induction and the equation $h(M) = h(M') + h(M'')$ we get

$$h(P \otimes_R M) \leq \beta(P)h(M') + \beta(P)h(M'') = \beta(P)h(M).$$

Remark 4.13 In [18] I also prove the Total Rank Conjecture in the following cases:

1. R is a "Roberts ring" (for example, a complete intersection ring) and char$(R/\mathrm{m}) \neq 2$ or
2. R is any local ring such that char$(R) = p$ for an odd prime p.

4.4.3 Matrix Factorizations

I provide no details, but would like to mention that in [5], Brown, Miller, Thompson and myself use cyclic Adams operations to prove a conjecture due to Hailong Dao concerning the vanishing of the theta invariant for isolated hypersurface singularities. Our proof requires developing a good notion of Adams operations for matrix factorizations. Since this involves $\mathbb{Z}/2$-graded complexes, the Adams operations of Gillet-Soulé cannot be used in this context.

4.4.4 A Generalization of the Total Rank Conjecture

Fix a local ring (R, m) and assume there exists a surjection $\pi : Q \twoheadrightarrow R$ with Q regular local, and let $I = \ker(\pi)$. Such a surjection exists if, for example, R is complete. For each finitely generated R-module M, choose a projective resolution $P^M \xrightarrow{\sim} M$ of M as a Q-module, obtaining a class

$$[P^M] \in K_0^{V(I)}(Q).$$

The assignment $M \mapsto [P_M]$ induces isomorphism

$$\rho : G_0(R) \cong K_0^{V(I)}(Q), \tag{4.13}$$

where $G_0(R)$ is the Grothendieck group of finitely generated R-modules.

We have a bi-linear pairing

$$\cap : K_0^m(R) \times G_0(R) \to \mathbb{Z},$$

called *cap product*, given on generators by

$$[P] \cap [M] = \chi([P], [M]) = \sum_i (-1)^i \operatorname{length}_R H_i(P \otimes_R M).$$

Using (4.13) this pairing induces the equivalent pairing

$$\tilde{\cap} : K_0^m(R) \times K_0^{V(I)}(Q) \to \mathbb{Z}.$$

An evident question is how this pairing relates to the two Adams operations on its source.

Lemma 4.5 *With the notation above, given an Adams operation ψ^k defined on the collection of all quasi-projective Q-schemes, we have*

$$\psi^k(\alpha) \tilde{\cap} \psi^k(\beta) = k^{\dim(Q)}(\alpha \tilde{\cap} \beta).$$

The proof of this Lemma is similar to the proof of [17, Theorem 7]. The basic idea is to reduce to the case where α lies in the image of the map $K_0^Z(Q) \to K_0^m(R)$ induced by π, for some closed subset Z such that $Z \cap \operatorname{Spec}(R) = \{m\}$. In this case, the equation holds from the projection formula. I omit the (complicated) details.

The following definition is a slight variation of a concept due to Kurano [13, §4].

Definition 4.8 (Kurano) With the notation above, an R-module M is called a *test module* (relative to π and ψ^k) if

1. M is a (f.g.) MCM R-module and
2. the class $[P^M] \in K_0^{V(I)}(Q)$ has weight $c := \dim(Q) - \dim(R)$ for ψ^k; that is, $[P^M] \in K_0^{V(I)}(Q)_{\mathbb{Q}}^{(c)}$ or equivalently

$$\psi^k([P^M]) = k^c[P^M] \text{ modulo torsion.}$$

Example 4.14 Suppose $R = Q/(f_1, \ldots, f_c)$ for some regular local ring Q and regular sequence of elements $f_1, \ldots, f_c$ in the maximal ideal of Q. Then R is a test module over itself (relative to the canonical surjection $Q \twoheadrightarrow R$) for any Adams operation. Indeed, it is MCM since R is Cohen-Macaulay, and the class of R in $K_0^{V(I)}(Q)$ is $[\operatorname{Kos}_Q(f_1, \ldots, f_c)]$, which, as we have seen, has weight c. (Note that $c = \dim(Q) - \dim(R)$.)

The following is equivalent to the (very strong) conjecture that every complete local domain has a (small) MCM module. See Kurano [13, §4].

Conjecture 4.3 Every complete, local domain has a test module.

Using these notions, I prove the following generalization of the Total Rank Conjecture.

Theorem 4.7 *Suppose $(R, \mathfrak{m})$ is a local ring of dimension d and* $\mathrm{char}(R/\mathfrak{m}) \neq 2$. *Assume $F \in \mathcal{P}^{\mathfrak{m}}(R)$ is a non-acyclic complex of the form*

$$F = (\cdots \rightarrow 0 \rightarrow F_d \rightarrow \cdots \rightarrow F_0 \rightarrow 0 \rightarrow \cdots).$$

If R admits a test module relative to ψ_{cy}^2 (for some surjection $Q \twoheadrightarrow R$), then $\mathrm{rank}_R(F) \geq 2^d$. *In particular, the Total Rank Conjecture holds for such a ring R.*

Remark 4.14 Let me make some comments about the complexes F appearing in the statement of this theorem: By the New Intersection Theorem [15], such complexes are the "narrowest" possible complexes of projective modules having non-zero, finite length homology. Moreover, when R is Cohen-Macaulay, any such complex is necessarily the resolution of a module of finite length and finite projective dimension, and thus the statement of this theorem is equivalent to the assertion of the Total Rank Conjecture in this case. Finally, I mention that the theorem would become false if F were allowed to be any "wider"; see [12].

Proof The second assertion follows from first since by the Auslander-Buchsbaum formula [3], the minimal free resolution of a non-zero R-module of finite length and finite projective dimension has the form of F.

Suppose $\pi: Q \twoheadrightarrow R$ is a surjection with kernel I such that Q is regular and that M is a test module for R relative to π and ψ_{cy}^2. As usual, let $P^M \xrightarrow{\sim} M$ be the minimal Q-free resolution of M.

As discussed above, we have a pairing

$$-\tilde{\cap}-: K_0^{\mathfrak{m}}(R) \times K_0^{V(I)}(Q) \rightarrow \mathbb{Z},$$

with the following two properties:

1. For any R-module N, if $P^N \xrightarrow{\sim} N$ is a Q-free resolution of N, then

$$[E]\tilde{\cap}[P^N] = \chi(E \otimes_R N) = \sum_i (-1)^i \mathrm{length}_R H_i(E \otimes_R N),$$

   for any $E \in \mathcal{P}^{\mathfrak{m}}(R)$.
2. For $E \in \mathcal{P}^{\mathfrak{m}}(R)$ and $P \in \mathcal{P}^{V(I)}(Q)$ we have $\psi_{cy}^2([E])\tilde{\cap}\psi_{cy}^2([P]) = 2^{\dim(Q)}[E]\tilde{\cap}[P]$.

Since, by assumption, we have $\psi_{cy}^2([P^M]) = 2^c[P^M]$, the second property yields

$$2^{\dim(Q)}[F]\tilde{\cap}[P^M] = \psi_{cy}^2([F])\tilde{\cap}\psi_{cy}^2([P^M]) = 2^c\psi_{cy}^2([F])\tilde{\cap}[P^M].$$

Using the first property, we deduce that

$$\psi_{cy}^2([F])\tilde{\cap}[P^M] = 2^{\dim(Q)-c}[F]\tilde{\cap}[P^M] = 2^{\dim(R)}\chi(F \otimes_R M).$$

The next part of the proof is very similar to the proof of Theorem 4.6 given above. In detail, we have

$$\psi_{cy}^2([F])\tilde{\cap}[P^M] = [S^2(F)]\tilde{\cap}[P^M] - [\Lambda^2(F)]\tilde{\cap}[P^M] = \chi(S^2(F) \otimes_R M) - \chi(\Lambda^2(F) \otimes_R M)$$

and the argument in that proof applies verbatim to show

$$\psi_{cy}^2([F])\tilde{\cap}[P^M] \leq \text{rank}_R(F)h(F \otimes_R M).$$

Combining these facts gives

$$2^{\dim(R)}\chi(F \otimes_R M) \leq \text{rank}_R(F)h(F \otimes_R M). \tag{4.14}$$

So far we have used neither that M is an MCM R-module nor that F is "narrow". These facts together imply that the homology of $F \otimes_R M$ is concentrated in degree 0:

$$H_i(F \otimes_R M) \cong \begin{cases} 0, & \text{if } i \neq 0 \text{ and} \\ H_0(F) \otimes_R M, & \text{if } i = 0. \end{cases}$$

In particular, $\chi(F \otimes_R M) = h(F \otimes_R M)$, and the Theorem follows from (4.14) by dividing by the positive integer $h(F \otimes_R M)$. □

References

1. J.F. Adams, Vector fields on spheres. Ann. Math. (2) **75**, 603–632 (1962)
2. M.F. Atiyah, *K-Theory*. Lecture Notes by D.W. Anderson (W. A. Benjamin, Inc., New York-Amsterdam, 1967)
3. M. Auslander, D.A. Buchsbaum, Homological dimension in Noetherian rings. Proc. Natl. Acad. Sci. USA **42**, 36–38 (1956)
4. M.K. Brown, C. Miller, P. Thompson, M.E. Walker, Cyclic Adams operations. J. Pure Appl. Algebra (2016). In press. Preprint. Available at arXiv:1601.05072
5. M.K. Brown, C. Miller, P. Thompson, M.E. Walker, Adams operations on matrix factorizations. Algebra Number Theory **11**(9), 2165–2192 (2017)
6. D.A. Buchsbaum, D. Eisenbud, Algebra structures for finite free resolutions, and some structure theorems for ideals of codimension 3. Am. J. Math. **99**(3), 447–485 (1977)
7. S.P. Dutta, M. Hochster, J.E. McLaughlin, Modules of finite projective dimension with negative intersection multiplicities. Invent. Math. **79**(2), 253–291 (1985)
8. W. Fulton, S. Lang, *Riemann-Roch algebra*, vol. 277 of *Grundlehren der Mathematischen Wissenschaften [Fundamental Principles of Mathematical Sciences]* (Springer, New York, 1985)

9. H. Gillet, C. Soulé, Intersection theory using Adams operations. Inventiones Mathematicae **90**, 243–277 (1987)
10. D.R. Grayson, Adams operations on higher K-theory. K-Theory **6**(2), 97–111 (1992)
11. R. Hartshorne, Algebraic vector bundles on projective spaces: a problem list. Topology **18**(2), 117–128 (1979)
12. S.B. Iyengar, M.E. Walker, Examples of finite free complexes of small rank and small homology. Acta Math. **221**(1), 143–158 (2018)
13. K. Kurano, Test modules to calculate Dutta multiplicities. J. Algebra **236**(1), 216–235 (2001)
14. P. Roberts, The vanishing of intersection multiplicities of perfect complexes. Bull. Am. Math. Soc. (N.S.) **13**(2), 127–130 (1985)
15. P. Roberts, Le théorème d'intersection. C. R. Acad. Sci. Paris Sér. I Math. **304**(7), 177–180 (1987)
16. Séminaire de Géométrie Algébrique du Bois-Marie 1966–1967 (SGA 6), Dirigé par P. Berthelot, A. Grothendieck et L. Illusie. Avec la collaboration de D. Ferrand, J. P. Jouanolou, O. Jussila, S. Kleiman, M. Raynaud et J. P. Serre, in *Théorie des intersections et théorème de Riemann-Roch*. Lecture Notes in Mathematics, Vol. 225 (Springer, Berlin, New York, 1971)
17. C. Soulé, Opérations en K-théorie algébrique. Can. J. Math. **37**(3), 488–550 (1985)
18. M.E. Walker, Total Betti numbers of modules of finite projective dimension. Ann. Math. (2) **186**(2), 641–646 (2017)
19. C.A. Weibel, *The K-Book*, vol. 145 of *Graduate Studies in Mathematics* (American Mathematical Society, Providence, RI, 2013). An Introduction to Algebraic K-Theory

LECTURE NOTES IN MATHEMATICS  Springer

Editors in Chief: J.-M. Morel, B. Teissier;

Editorial Policy

1. Lecture Notes aim to report new developments in all areas of mathematics and their applications – quickly, informally and at a high level. Mathematical texts analysing new developments in modelling and numerical simulation are welcome.

   Manuscripts should be reasonably self-contained and rounded off. Thus they may, and often will, present not only results of the author but also related work by other people. They may be based on specialised lecture courses. Furthermore, the manuscripts should provide sufficient motivation, examples and applications. This clearly distinguishes Lecture Notes from journal articles or technical reports which normally are very concise. Articles intended for a journal but too long to be accepted by most journals, usually do not have this "lecture notes" character. For similar reasons it is unusual for doctoral theses to be accepted for the Lecture Notes series, though habilitation theses may be appropriate.

2. Besides monographs, multi-author manuscripts resulting from SUMMER SCHOOLS or similar INTENSIVE COURSES are welcome, provided their objective was held to present an active mathematical topic to an audience at the beginning or intermediate graduate level (a list of participants should be provided).

   The resulting manuscript should not be just a collection of course notes, but should require advance planning and coordination among the main lecturers. The subject matter should dictate the structure of the book. This structure should be motivated and explained in a scientific introduction, and the notation, references, index and formulation of results should be, if possible, unified by the editors. Each contribution should have an abstract and an introduction referring to the other contributions. In other words, more preparatory work must go into a multi-authored volume than simply assembling a disparate collection of papers, communicated at the event.

3. Manuscripts should be submitted either online at www.editorialmanager.com/lnm to Springer's mathematics editorial in Heidelberg, or electronically to one of the series editors. Authors should be aware that incomplete or insufficiently close-to-final manuscripts almost always result in longer refereeing times and nevertheless unclear referees' recommendations, making further refereeing of a final draft necessary. The strict minimum amount of material that will be considered should include a detailed outline describing the planned contents of each chapter, a bibliography and several sample chapters. Parallel submission of a manuscript to another publisher while under consideration for LNM is not acceptable and can lead to rejection.

4. In general, **monographs** will be sent out to at least 2 external referees for evaluation.

   A final decision to publish can be made only on the basis of the complete manuscript, however a refereeing process leading to a preliminary decision can be based on a pre-final or incomplete manuscript.

   Volume Editors of **multi-author works** are expected to arrange for the refereeing, to the usual scientific standards, of the individual contributions. If the resulting reports can be forwarded to the LNM Editorial Board, this is very helpful. If no reports are forwarded or if other questions remain unclear in respect of homogeneity etc, the series editors may wish to consult external referees for an overall evaluation of the volume.

5. Manuscripts should in general be submitted in English. Final manuscripts should contain at least 100 pages of mathematical text and should always include

   - a table of contents;
   - an informative introduction, with adequate motivation and perhaps some historical remarks: it should be accessible to a reader not intimately familiar with the topic treated;
   - a subject index: as a rule this is genuinely helpful for the reader.
   - For evaluation purposes, manuscripts should be submitted as pdf files.

6. Careful preparation of the manuscripts will help keep production time short besides ensuring satisfactory appearance of the finished book in print and online. After acceptance of the manuscript authors will be asked to prepare the final LaTeX source files (see LaTeX templates online: https://www.springer.com/gb/authors-editors/book-authors-editors/manuscriptpreparation/5636) plus the corresponding pdf- or zipped ps-file. The LaTeX source files are essential for producing the full-text online version of the book, see http://link.springer.com/bookseries/304 for the existing online volumes of LNM). The technical production of a Lecture Notes volume takes approximately 12 weeks. Additional instructions, if necessary, are available on request from lnm@springer.com.

7. Authors receive a total of 30 free copies of their volume and free access to their book on SpringerLink, but no royalties. They are entitled to a discount of 33.3 % on the price of Springer books purchased for their personal use, if ordering directly from Springer.

8. Commitment to publish is made by a *Publishing Agreement*; contributing authors of multiauthor books are requested to sign a *Consent to Publish form*. Springer-Verlag registers the copyright for each volume. Authors are free to reuse material contained in their LNM volumes in later publications: a brief written (or e-mail) request for formal permission is sufficient.

Addresses:
Professor Jean-Michel Morel, CMLA, École Normale Supérieure de Cachan, France
E-mail: moreljeanmichel@gmail.com

Professor Bernard Teissier, Equipe Géométrie et Dynamique,
Institut de Mathématiques de Jussieu – Paris Rive Gauche, Paris, France
E-mail: bernard.teissier@imj-prg.fr

Springer: Ute McCrory, Mathematics, Heidelberg, Germany,
E-mail: lnm@springer.com

Printed in the United States
By Bookmasters